SOFTWARE DEV HAS STALLED

Advancement in Software development approaches have stalled.

Pace is a refreshingly pragmatic guide to overcoming the shortcoming of today's software development.

Littered with decaying practices, it is a far cry from what it was 20 years ago, when the Agile revolutionaries rescued it from Waterfall. Once at the forefront, Manifesto in one hand and Scrum certification in another, these revolutionaries have quietened down.

Over the last 20 years, software has undergone massive changes—such as the introduction of apps, smartphones, SAAS, and online gaming, as well as its inclusion as a critical part of vehicles' value proposition. During this 20-year period the first fighter aircraft (F-35 Lightening II) to have more involvement in developing the software component, than the hardware was planned, and built. Considering the fact that modern software development is all about flexibility and striving for improvement, how far have we actually come in the same 20 years?

In Pace, Peter Cronin captures stories from many of the projects he's worked on, letting these serve as teaching ground for his breakthrough software development approach—Pace. Peter doesn't shy away from slaying a few sacred cows either, such as doing away with sprints and deadlines.

Peter explains that Pace ultimately comes down to a few key rules.

1. Find and optimize performance of the bottleneck in your development cycle
2. Protect this optimized performance from change and variation
3. Accelerate development while increasing quality

Conceptually simple. Complex to execute.

Pace then looks into the management decision making techniques and proposes that—although you can benefit from developing high-quality code faster and faster—this benefit will be lost if managers can't keep up by making robust decisions quickly.

Development cannot be siloed from the rest of the organisation, and indeed we are well overdue for an approach to development that does not optimise development at expense of the commercial reality of the business.

Pace is intended for:

- lovers of Agile looking for the next big step,
- those who don't love Agile and are looking for an alternative that will deliver on the benefits Agile promised all those years ago, AND
- founders struggling to scale (and keep the magic).

PACE

PETER CRONIN

Pace.

Published by Viago International. Printed by The Book Printing Company Australia. Paperback Edition. 2020.

First Edition.

www.viagointernational.com

CONTENTS

SHOUT OUTS!

Pace has been built by drawing on areas of expertise, techniques, and concepts from an extremely wide range of areas and sources. Many of those too minor to note, or to remember the small role they play in the bigger picture. There are also a few that have played a significant role, which I acknowledge here. I try not to specifically align myself with any methodology or concept as I often see people who do are putting a metaphorical flag in the ground, some with a degree of religious fervour. This leads to the same people defending their flag rather than being open to the best possible solution. It is only with this openness supported by the assumption techniques that Pace could have come about and been so applicable to different teams and businesses while also being so effective.

Agile

The concepts outlined in the Agile Manifesto are extremely relevant for software businesses, and how they manage the pressures and counter pressures defines their success. Meanwhile the mechanisms to be Agile; Scrum, Sprints, SAFe, etc have provided structures that are immensely valuable to software development everywhere. As well as this they have formed much of the groundwork on which Pace has been developed, without which Pace couldn't be what it is.

Theory of Constraints

The importance of identifying and managing constraints (we call bottlenecks in Pace) has been essential to both the effectiveness of the solution, and the relative simplicity. The constraint approach allows for solutions which are robust in the principles that are applied, and specific states that must be true in teams and businesses, however rather flexible in the approach of how to achieve those states. This has given Pace the ability to be both more effective than the incumbent methods in software development, while also more applicable to a wide range of software companies.

PACE

PETER CRONIN

PART 1

A DAY IN THE LIFE

CHAPTER 1

Dave's Fiasco

Dave doesn't know it yet, but he's about to have "one of those days"...

12:15pm—He's been at work less than an hour, and he's already dealing with an emergency. Sarah, one of the product owners, is on annual leave this week; however, she neglected to tell everyone (or just forgot) that a bug fix was promised to a major client by Monday. Unfortunately, it's now Tuesday morning and the bug fix wasn't planned into the sprint. Fifteen minutes ago, Dave took a frantic call from Derek in customer service, demanding to know what was going on.

"I have the customer waiting on the phone right now, and they're NOT happy!"

Dave begrudgingly promised to look into it immediately.

"I have absolutely no idea what's going on—it's Sarah's job to deal with customers, not mine," Dave told Ritika, a senior developer. "And I don't have the capacity to add more work—I have a backlog of my own bug fixes that I need to deal with today."

"Well, if we don't expedite this one, Sarah's gonna rip into you when she's back," Ritika said. "Also, we could lose the customer. They're very upset—apparently, they expected the patch to be deployed first thing this morning. Until Sarah gets back, you're going to have to deal with this."

Dave sighed, imagining the ticket was not even properly detailed in the system. "Okay, let me try and find it."

1:30pm—The emergency is finally dealt with, but then an urgent email arrives just as Dave has settled in with a cup of coffee. Before he has a chance to start catching up on his Slack messages, he's interrupted by an email alert from Derek.

"An urgent feature request has just come through from sales. Can you get the design spec'd today so I can get the sign-off from the customer? Jenny over in sales promised we could get the design sorted today. If we drop the ball on this one, it won't look good."

Dave put down his cup of coffee, now cooling down, and wished they would just finish the work that had been started first. But that wasn't the company's way. Their unofficial motto was "Just get it done." It seemed as though both his coffee and the Slack messages were going to have to wait.

5:25pm—Dave has his fingers crossed; the last half hour has been free of interruptions. He's been able to specify some urgent designs and actually get some coding done. If he can keep this up, he might actually finish work almost on time today. He's about to move to the next thing on his list when the phone rings, ruining his productive streak—it's the accounts department.

"We have a minor invoicing crisis," Henry from accounts said brusquely. "I need your help to sort out a mix-up."

Dave groaned. He could tell from Henry's tone that this would consume the rest of his afternoon. "Mix-up? Did the team track their time improperly again? Never mind...who's the customer? I'll be right there."

7:45pm—Dave's been putting out fires all day. He can't find time to get the urgent tasks processed, much less the non-urgent ones. Ritika dropped by his desk on her way out the door.

"Have you been able to review Sam's code yet?" she asked.

"You've got to be kidding," Dave answered without turning around. "I've had interruption after interruption since I arrived today. I've been putting out fires left, right, and centre. I can't find time to get the urgent designs spec'd, much less do reviews."

Ritika nodded sympathetically. "Can you talk to Chris about it? Maybe there's something that can be done to ease the load."

CHAPTER 2

How We Got Here

The Waterfall model was believed to be the first software development process model to be introduced to the software engineering industry. This belief is supported by the timeline of events surrounding the emergence of the Waterfall model; the term 'software engineering' was coined in 1965 by Margaret Hamilton, a developer at NASA, only five years before the first publication about the Waterfall approach.

The first formal description of the Waterfall model in the software industry was outlined in 1970, in a paper called "Managing the Development of Large Software Systems," published by Winston Royce. Royce outlined the model using a graph, which looked very much like a waterfall going from each step: system requirements, analysis, coding, testing, and operations. The term 'waterfall' wasn't used until 1976.

The software industry before and during the 1970s was very different to what it is today. Most software projects were large software packages, often coupled to infrastructure, and so were managed and developed in the same way—applying infrastructure-based (hardware) development methods to the creation of software. Some examples include government projects and large projects for NASA, such as the software development that went into the rockets to take Man to the Moon.

Due to the rigidity of specifications for those projects, there was little room for variation in the process. Any changes made to the specifications during development would have resulted in high costs, so the very rigid and structured approach of Waterfall worked well.

Benefits of Waterfall Software Development

Despite different technology, the goals of a software project back in the 1970s were much the same as they are today; deliver projects on time, meet budget, and meet scope. With the highest costs of a software project occurring in the coding and testing phases, the Waterfall model allowed developers to reduce coding and test time and cost, while still delivering full scope by analysing and capturing a strict documentation of requirements before any planning or development occurred.

The Waterfall model then ensured completeness of a project, and all requirements to specification, as each stage in the process needed to be completed before moving on to the next one.

The Challenge

When people started using software more and more for business, there was a rise in the amount of business software that was starting to be developed. The commercial software was being developed for businesses initially, and then started to come out for end users as well. Software developers of these modern commercial products found the traditional processes were too heavy on the infrastructure.

Software product development using the Waterfall approach created a number of problems:

Too much documentation upfront

- Spending time on and committing to things that are likely to change as the project progresses.

Contract negotiation with customers designing big plans

- A focus on us vs them, giving customers 'proof' of milestone completion, rather than being open with information and working with the customer to create more effective software as the project progressed.

Adherence (or lack thereof) to the designed plans

- Developing modules and features that are no longer appropriate because the plan said so, or ignoring the direction entirely because the plans are no longer relevant.

Focus on processes and tools rather than on the developers trying to implement these

- Developers must comply with processes mandated by those with no experience in software projects, focused instead on project reports and milestones, rather than supporting development practices.

Long coding runs which lead to increased lead times and increased resistance to rework

- Developers can spend weeks coding without realigning with the team or having the direction checked with a team leader.

Long periods of time between review so errors propagated throughout the product

- Chaos caused by leaving reviews to be done together, and integrations after that would often be so complicated to fix that they would be patched back together, never really being fixed.

Agile: The New Approach

Agile software development was born out of the need to address these challenges created by the traditional infrastructure-style Waterfall processes. Software developers identified that it was far more useful to have a process that could respond to changes in customer requests, and to get something basic working quickly, then adjusting and iterating from there. Using an Agile approach, developers could adjust the software based on their experience with the previously developed chunk of the software and move forward using an iterative process.

The emergence of the Agile approach enabled software companies to more quickly develop software that created value for customers. It also enabled software development teams to be more responsive to changes and reduce the amount of rework required due to changing specifications.

Faster delivery of a product to market or value to the customer.

A basic, functioning piece of software is developed quickly to increase speed to market or value to the customer quickly. The value transfer to the customer is much sooner than in a traditional software project.

More adaptive and flexible.

The working software is iterated using feedback on how the customer uses it or feedback about what more they would like from it. Customers don't need to make all the decisions about the final product up front—they can make decisions, changes, or adjustments based on what they discover from using the early versions of the product. Software is more aligned and tuned with customers' needs based on what is discovered during iterations.

Many smaller and more frequent reviews (leading to higher-quality software with fewer bugs).

Developers complete smaller chunks of code, so reviews occur more frequently, and take less time to do. Errors or bugs in the code are caught early on, which minimizes the amount of rework that needs to be done.

Better communication (creating a more holistic approach overall).

People and interactions are favoured more than processes and tools under the Agile Manifesto. There are more frequent communications with the customer throughout the entire development process. It is much more end-user focused. There is also more communication internally within the cross-functional software teams responsible for each part of the product. Communication occurs between the whole team throughout all stages of the process.

CHAPTER 3

What Is Agile?

There are two primary ways Agile is defined:

1. Agile theory (from the Agile Manifesto)
    a. Individuals and interactions over processes and tools
    b. Working software over comprehensive documentation
    c. Customer collaboration over contract negotiation
    d. Responding to change over following a plan

    That is, while there is value in the items on the right, we value the items on the left more.[1]

    This definition is useful for discerning intent, because it's the original source of Agile. The Manifesto focuses on Agile practices, as opposed to Waterfall or traditional project management practices.
2. Agile in practice. What are people actually doing and implementing? This definition is far more useful, because while theory is nice, it doesn't translate well to the day-to-day practices of businesses and development teams.

In practice, what most companies are doing is primarily Scrum or sprints. If you ask the owner or senior manager of a small or medium-sized software company, "Do you operate using agile principles?" or, more colloquially, "Is your company Agile?" they would most likely base their answer off whether or not their development teams are using sprints to plan, coordinate, and execute their work. The term has actually become a trend or buzzword for companies which want to be perceived as dynamic or

1. https://agilemanifesto.org/

progressive. In this sense, the "dictionary definition" of Agile has changed over time, much like many other words in the dictionary, as people use the word differently in theory than in practice.

So, how are people actually practicing Agile? Software teams are putting the theory into practice via Scrum methods, especially sprints. This involves both the sprint planning meetings and the actual running of the sprints themselves. We'll dig more into sprints in the remainder of this section.

Planning Sprints

The first decision development teams need to make is how long they want the sprint to be. Most sprints, especially at small or medium-sized software companies, are two weeks in length. Although one-week sprints are recommended, many companies find these quite difficult to maintain. I suppose you could say that the shorter a sprint is, the more Agile it is, but shorter sprints tend to have more negative side effects than positives. Negative effects are going to be more minor during longer length sprints, and more extreme during shorter sprints—which is probably why companies don't do one-day sprints.

So, most companies settle in at fortnight-length sprints, as these seem to be the ideal length for manageability. The planning stage of these sprints is extremely important for the success and production levels of the sprint. The whole team needs to get together to plan the work that will be done over the length of the sprint. The work usually includes a certain amount of feature work or development work, or services. Most software companies will also fit in some level of bug fixes, whether working on services or products.

When these sprints are planned, people are essentially estimating how much work they can get done during that time period, and whether any leftover work then flows over into the next sprint, or whether the team would need to spend a few late nights towards the end of the sprint to cram it all in. The approach varies from company to company; some shoot for very tight and accurate estimates. However, this uses the dangerous assumption of the best-case scenario, which, of course doesn't usually eventuate. In that case, the work would roll over into the next sprint, affecting the next sprint's estimation and planning process.

Sprint planning is generally based on the capacity of each person on the team, whether that's 40 hours or 60 hours per week, per developer, for the length of the sprint. The type of work is then planned into that capacity, and will vary depending on how much the team can realistically get done during the sprint. The work might come from a backlog of projects that need to be finished, from requested service work or features, from bug fixes that have been requested from customer service, or a mix of all of these.

In addition to the pre-sprint planning process, there's also a level of daily planning that occurs during the sprint in the form of daily meetings. Some people call these daily Scrums, or daily standups. The term "standup" nicely reinforces the fact that these daily planning meetings should be brief and to the point. These planning meetings are a great opportunity to ensure the team is on track, and make everyone aware of the pace. These standups can become drawn-out and repetitive if teams aren't careful, so it's important to keep these types of planning meetings short and sweet.

Once the sprint is planned out, the work is then released to the team to execute.

Executing Sprints

As previously stated, sprints have all hands on deck, with all development team members working at full capacity for the length of the sprint. Sometimes the capacity of team members ends up being overestimated, forgetting the fact that some of the team's resources may gate the productivity of other resources on the team, because the amount of handovers between team members may increase "dead" or waiting time during the sprint.

For example, one developer may be waiting for something to come from another team member, in order to stick to the plan of the sprint's workload. While waiting, they may: A) sit around doing nothing, endangering the productivity of the sprint; B) pester the person holding up the project, annoying an already hard-working team member; C) drag out the task they're currently working on while waiting, again, not working to their potential; D) pick up something else off the pile and start on a different task, which can slow the sprint's progress even further if that new task is outside the scope of the sprint. None of these are great options, which is why it's important to factor waiting time and handover time into the capacity of the sprint team.

Good teams with successful sprints will have good communication and teamwork, so they can smoothly transfer tasks from one person to the next in the process. Good oral and written communication is bolstered by some sort of task-tracking process so that all team members can know where each item stands in the sprint. If someone is simply mentally tracking and pacing their own work, that doesn't help their team member who needs to know when they will receive the handover for that particular project. If, while executing the sprint, team members aren't clear about transitions between tasks, it can burn precious sprint time and cause the work to creep up, resulting in a stressful burnout at the end of the sprint.

One of the biggest enemies of the Agile sprint is the interruption. The longer the sprint, the more likely interruptions are to occur. These interruptions may come from customers requesting new bug fixes, or other urgent or semi-urgent requests. Your sprint team will have to decide whether there is time to interrupt the current sprint work to fulfill new tasks without throwing off the whole point of the sprint, or whether your customers will have to wait until the sprint is over. Perhaps you will need to alert your customers that all non-emergency tasks will need to wait until a certain date, so they stay satisfied until their work can have your full attention. Some teams factor a buffer into their sprints, for cases like this. It leaves a little extra padding for things that come up, so that interruptions don't derail the entire progress of the sprint, by forcing the team to throw out an integral part of the sprint work. Again, the most helpful way to battle evil interruptions is with good teamwork and collaboration.

Throughout the sprint, there will be the previously discussed standup meetings and check-ins from the team leaders, to make sure that the team is on target to meet the sprint deadline. These standups are important, because even with great sprint visibility tools (discussed in the next section) it's difficult to tell where exactly the progress on each task stands.

Visibility

An important part in the sprint execution is the company's method of visibility. Once the planning is completed, there's usually some sort of visual method of showing the work to be done during the sprint. It might be some sort of Kanban or task card/Post-It board to track the progress of work. These visuals will usually have similar component categories:

- Backlog (things that haven't been started on yet)
- Doing (tasks that people are currently working on)
- Done
- (sometimes there will be an extra column for Testing)

In some cases, these visibility boards will be manual; in other cases, digital-based in Jira—or whatever software platform the team is using to track the sprint's progress. And in some cases, teams don't really use a task management system; they might work from an Excel list, or just have developers track their own work instead of collaborating. Some might use a whiteboard or be colour-coded with a different colour for each team member or project. Although there's a very broad range of visual application, most commonly sprints are tracked on a Kanban board, with tasks passing from person to person. Most companies prefer to have a hands-on, physical board to work on, even if they're also using a digital task or ticketing system.

Even with visual tools like this, it may be hard to track sprint progress. The Doing column may have multiple tasks shown for any particular individual, so it can be hard to know exactly what each developer is working on at any one given time. It's also difficult to see that progress relative to the burndown of time over the length of the sprint. When team members end up spending a lot of time answering questions that should be addressed by the visibility tools, it is frustrating to those developers who feel like they're spending precious time updating team leaders instead of actually working on the tasks at hand.

Visibility can also be a multitasking headache. There's a mental burden of trying to decipher the task board, mentally switch between tasks, and understand what needs to be done overall.

Sprint Conflicts

Teams running sprints often encounter a number of conflicts stemming from the above stages of Planning, Executing, and Visibility. Some of these have already been briefly touched on, but all should be taken into consideration when companies "go Agile."

Sprint Length

This is the conflict that requires the earliest resolution in the Scrum implementation process. Once this conflict is settled, it doesn't often change; once the sprint length is decided upon, it's fairly set in stone. Generally, shorter sprints involve more (and more frequent) planning and deciding what work should happen, which decreases the amount of time actually spent "sprinting" and writing code. Longer sprints have the potential for unknown variables and wasted effort on team output that doesn't turn out to be useful. The longer the sprint, the more issues with traditional project management will pop up; the shorter the sprint, the more the benefits of sprinting are felt (and the difficulties that come with it). Teams with a "process of ongoing improvement mentality" often choose a shorter one-week sprint. Those that find that too gruelling will choose a more comfortable two-week sprint. We'll further discuss the importance of iteration length (sprinting vs. moving at pace) as a critical element in a later section.

Planning for Capacity vs. Outcome

Agile theory maintains that teams should be focusing on an outcome—and many top dogs in the industry do—but trying to plan work around a specific deliverable doesn't fit into an abstractly chosen two-week deadline as neatly as you'd hope. Who would have thought that modern-day employment standards, based on the speed at which the Earth rotates, crossed with a week-based calendar designed over a thousand years ago, based on the waxing and waning of the Moon, doesn't fit with the time it takes for a team of people to create a meaningful chunk of software?

This means that the work is either jammed to fit into two weeks, stressing out and overloading the entire team, or expanded to fill two weeks, meaning there will be team members who aren't loaded up to full capacity.

Instead, the plan is based on each individual's capacity, but in doing that you'll still need to account for certain expert resources in the team who constrain the total team's output (more on this later). This means the other team members are either unable to get all their work done, or in their efforts to be efficient they're limiting the output of the expert resources, and therefore the team. This doesn't even take into consideration all of the Grey Time[2] the team has in their handovers, or completion of work in general.

Ultimately, this ends with undue stresses on the team, because they aren't meeting the sprint deadlines and targets. This conflict will tip the balance based on whether or not people have the general ability to put aside efficiency metrics to deliver outcomes. Everyone being extremely busy isn't a good thing—there's a reason that emergency services have spare capacity instead of being measured by their "productive" time. In software, this planning balance will generally tip towards the outcome in smaller start-ups and toward "busyness" in larger businesses.

Handling Outside Work

Another conflict is how to handle work not being done within the sprint. Of course, one way to deal with this is to make everyone on the team have a stressful "eleventh hour" before the end of the sprint, cramming all the work in, no matter what. Developers are far from stupid—you don't have to be particularly intelligent to know that you can always gain time by cutting corners. In order to make sure everything gets finished during the sprint, they could drop quality reviews or shorten them to a quick skim, but the consequence of this would be extra rework to deal with later. These cut corners and rework may be caught by the team, or worse, by the customer in a few weeks when they need a bug fix. This creates a "snowball effect," with the technical debt piling up more and more.

With the resulting rework, the unfinished sprint tasks would need to be rolled over into the next sprint, which only goes to show the team members that deadlines are not firm and absolute. The prevailing team attitude toward these "deadlines" becomes lackadaisical, because they're just paying lip service to a work tracking process that doesn't really mean anything to them, unless someone steps in and firmly implements abso-

2. See end of chapter for explanation of Grey Time

lute deadlines as they should be.

The dreaded "unplanned" creates an erratic decision-making process within the planning and execution of sprints. The team may be sprinting away, when "that" manager appears, and the whole team inwardly (or outwardly) groans. That manager might be from sales and has just promised some great new feature, customization, or integration to land an important new sale. That manager may be from customer service and is sick of dealing with an unhappy customer yelling over the phone. Or worse, that manager could be the founder of the company who just wants to make his dearest, oldest customer happy by any means possible. Either way, the interruption eliminates any hope of completing the sprint on time and on spec; something will have to give, and the turbulence generated by putting down half-completed work to be picked up again at a later date will never be recovered by the team.

Some teams have the freedom to decide the outcome of this conflict themselves; others are at the mercy of another member of the business. Either way, the decision needs to be made whether or not the team will shove additional work into the sprint, causing interruptions, which in turn increases multitasking and time allotted per task. Ideally, the decision maker would realize the impact, and refrain from shoving interruptions into a sprint with "urgent" requests from "important" customers. Insisting on this outside work would only raise stress levels of "that" manager, and breed resentment within the dev team, circling back around to the initial option of cramming the work in at any cost. Going back to Dave's fiasco, the goal is to not make the company's motto "Just get it done."

Bug Fixes

Teams have more control than managers over bug fixes in sprints. When there is a higher volume of bugs to squish in the sprint, perhaps because there were more serious level tickets raised during the previous few weeks, it displaces "business as usual"—whether that's new features, customer developments, or roadmap items—which puts pressure on the business' progress on the product. Slowing of progress is a dangerous slope, as there are many important things that can chip away at time spent until the team ends up making very little progress at all, and the market starts to catch up, or even leave them behind. These pressures can present themselves in different ways—from investors to sales, management roadmap milestone bonuses, etc.—but they all lead to increasing time spent on progress work and decreasing time spent on bug fixes. Fewer bug fixes mean there will be more bugs, and more bugs make the users mad, which means the next sprint will need to include more bug fixing.

Poor Visibility

As previously mentioned, all sprints use some sort of visibility tools in their planning, whether that's digital, or a physical task board. Referred to as Kanban boards by most software companies, they will show the planned tasks for a sprint and their current status broken across categories. Depending on the level of detail in this task board, visibility can cause a higher or lower level of dilemma for development teams during a sprint. If the board has very detailed information—such as updated estimates, tasks being actively worked on, a burndown chart or other progress vs. planning indication—it offers much more assistance and effectiveness in sprint execution. However, along with that level of detail comes the need to update the board much more frequently. Sometimes, these updates become a frustrating admin burden which interrupts other, more productive, activity. This leads to less quality development time, and if the team doesn't update as necessary, leads to decreased visibility that makes it more difficult for the team to deliver the work at a smooth, consistent pace.

Grey Time

Many of these conflicts are exacerbated by the mishandling of Grey Time.

Whenever you allot a time amount to a task, you need to account for "Grey Time." This includes many situations such as information not in the system, details that require clarification, or handovers between team members.

For example, Sophie might say it will take her 30 minutes to do a particular task, so you'd calculate that she can do 16 of those tasks per day (8 hours of work time divided by 30 minutes each). Sophie will almost certainly say, "No way—I can do 10 per day!" (she wasn't accounting for Grey Time in her 30-minutes estimate). This Grey Time for Sophie may include a handover from the previous team member in the task process, who was late because they took time to tell Sophie about the project on their way to lunch. Or Sophie chatting with Jim about the new Jaguar parked outside, after they had a coffee break in the kitchen.

Grey Time is neither good nor bad. Some things just happen; others happen because we're humans, not robots; some shouldn't happen and can be improved upon in various ways (some discussed in this book). Grey Time is simply an important consideration in the planning and management process of sprints. This means estimations of it need to be built into standard operating procedures, giving customers estimations of bug fix completions, etc. Grey Time must be accounted for, in order for the plans of the sprint to be successful.

These common sprint conflicts, and others, take place in software teams around the world; some dilemmas may occur daily, some monthly, and some only once, but each team and company need to figure out its own balancing act. This balancing act is individualized to each company, and may take many revolutions of the "figure-eight" loop to get there. The figure-eight means doing less of X which gives us less Y, so we do more X to get more Y, accidentally lowering the Z we get, which leads us back to doing less X, and so on…

A couple of diagrams demonstrating the figure-eight loop:

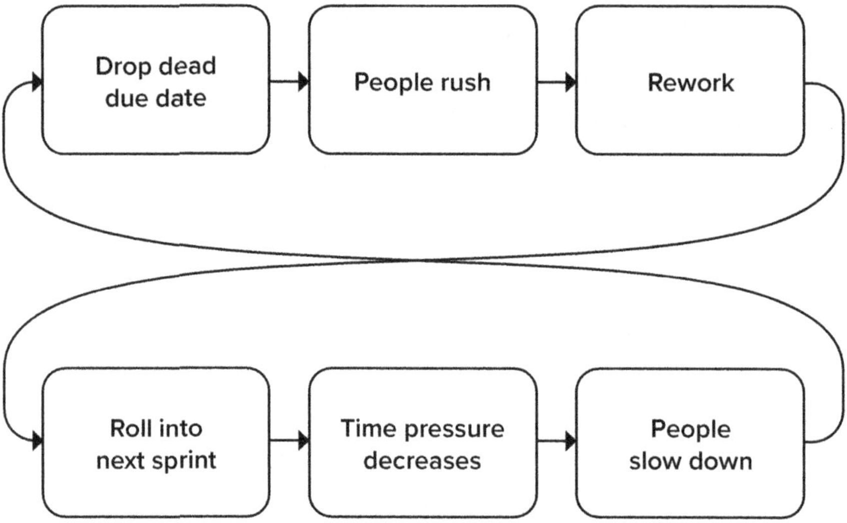

Fig 1

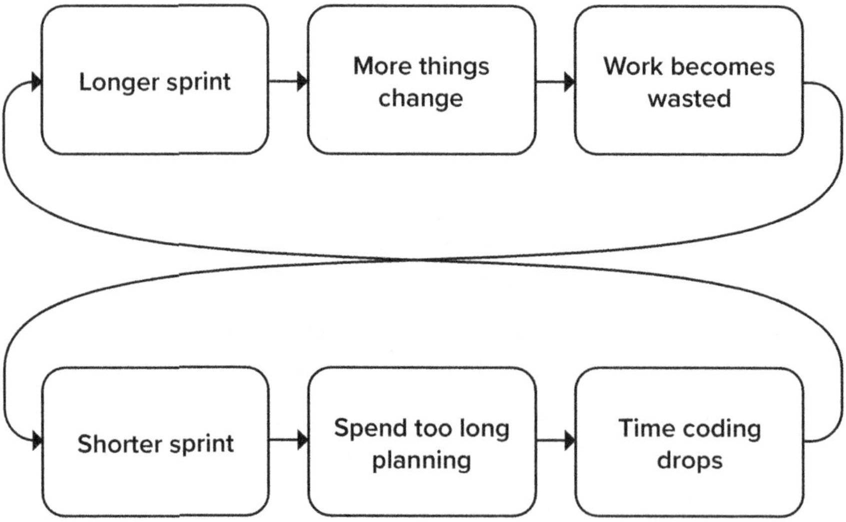

Fig 2

These dilemmas have stalled the progress of Agile for many. Sprints were once the way forward to more effectively implement the left side of the four 'over' statements from the manifesto. While they have done this effectively, the majority of software teams are now in the current reality where Sprints and Scrum synonymous with 'Agile' in the business.

The core of the manifesto is the four 'over' statements, which are capped with the statement, "while there is value in the items on the right, we value the items on the left more." In many cases, this has been understood to mean, "always do the items on the left", and in worst cases the manifesto is used as a hammer to beat people up when they aren't being as Agile as someone would like. In reality, the items on the right side of the manifesto represent commercial drivers that are also of value (as stated by the authors), which means people in the business with a strong commercial focus—such as sales or upper management—will push back on the left and move forward in support of items presented on the right-hand side. This leads to arguments over actions on the left or actions on the right, rather than on a focus of trying to achieve both. The original text of the manifesto should be interpreted to emphasize that both are critical, but in the event of a true deadlock, the left side should be favoured.

Scrum has brought us all the way to this point, and it now sits at an impasse—it now represents the status quo once maintained by Waterfall. Scrum, and more specifically sprints, are now too slow. Their lack of progression has stalled Agile in many software teams and leaves them ripe for disruption from a more Agile approach—one that favours faster-moving, smaller iterations of work that are managed in such a way as to allow for continuous development, and are planned on true development needs and estimates, not shoe-horned into short project timeframes of 2 or 4 weeks. The dilemmas above outline the equivalent of the 'over' statements in a team running sprints.

As you may suspect, the point of identifying these sprint conflicts is not to pick one side at the expense of the other, but to find a way to attain both. Software teams and businesses often face the dilemma of speed vs. quality. Finding ways to get both at the same time is how progress is made in software development—this is what Agile methods like Scrum deliver in comparison to traditional project management. Later, we'll explore application methods in your own business, but first we'll introduce you to Pace, which delivers more speed and quality by resolving the issues present in sprints.

CHAPTER 4

Chris' Dilemma

Remember Dave? If you felt sorry for him, make sure you also spare a thought for his boss, Chris. Chris is the hardworking owner of the company Xyphyr. He works long hours, and so does his team. In almost any other industry, they would all be reaping rich rewards; however, the need to scale to make money in today's software companies coupled with rising competition and never-ending customer demands constantly pressure his cash flow, making it a constant headache to attempt to turn a profit. Chris thinks, *We have no other options—we either drop prices, or we increase the feature offering to match what other companies are doing!*

In addition to dealing with unexpected emergencies, Xyphyr also has to jump through hoops to win new customers. Not only do those new customers want bug-free software, but they also demand ever-increasing amounts of new features and customization. One customer demands tailored calendar integration for their HR processes. Another wants specialized email tracking. And yet another demands accounting integration from an obscure company.

This not only adds to the cost of doing business (placing even more pressure on Chris' margins), it also increases the pressure on Chris' already overworked development team. And the frustrating thing is, no matter how much "hoop jumping" Chris does, by expediting a new feature the customer desperately needs, his clients show little gratitude. They'll probably be back on the phone in a week, demanding another bug fix. *If only I could find the time to get on top of the direction of the product and get ahead,* he thinks to himself.

It's this kind of persistent problem that is currently the subject of a heated discussion between Chris and Dave. Everyone else has finally gone home.

"We need this new feature," Chris stressed to Dave. "Desperately."

"Chris, I'm telling you—development has absolutely no capacity to deliver it. With Sarah away, I was completely overwhelmed today. I've been hoping to get a chance to design that new feature for the customer that Jenny won the other day, but instead spent most of the day putting out fires. It's not going to be much better, even when Sarah gets back."

"Well, that will change once Jenny starts losing clients because we can't match what InstaHR or FaceHR are now offering to every client," Chris pointed out.

"Discount again?" Dave suggested helplessly.

"If you look at the P&L for the last six months," Chris said, waving a hand at his monitor, "there's absolutely no fat left on our bottom line to discount any further." Chris sighed. "We have two tools in our kit—a hammer and a screwdriver. That's all. We have no other options—we either drop prices or we promise customizations to get ahead of what others are doing. And we can't do any better on price than we're currently doing. If we do nothing, we'll bleed the regular customers who keep us even marginally profitable. And perhaps if things go well, we'll be able to add more developers to cope with any increased backlog. But there's no way I can afford to add anyone else right now."

Dave was lost for words. He knew that it was wrong to add yet more customizations to her already overworked development team. Intuitively, he felt that it would be a disaster—further increasing already slow delivery times, adding to the number of critical bugs and near misses, and leading to more and more unhappy clients. And the way reputation tended to spread online, it would make it harder to win clients in the long run, not easier!

He also knew that, even though Chris understood exactly what Dave meant, he would never convince him that offering the extra features or customizations was actually counterproductive. They were simply stuck between a rock and a hard place. As he left for the night, Dave wondered how on earth she was going to explain this to the team in the morning.

Meanwhile, Chris disconnected his laptop from his monitors and slid it into his leather satchel. Swinging it over his shoulder and heading for the car, he couldn't help but once again think about how it was supposed to be different by now. He and his co-founder often discussed their vision, now feeling more like a blurry dream, of how they would have teams that look after themselves while he focuses his time on the strategy of the product. He should be meeting with customers, planning the future features pro-actively rather than reacting to demands, and handing over a high-level roadmap to people like Dave to get on with. He should be utilizing his strengths in dev design, not building an unfortunate skill in bullish project management style techniques—like threatening people with due dates.

PART 2

DEVELOPMENT AT PACE

CHAPTER 5

Introduction to Bottleneck Management

Systems with a series of tasks that must be passed from one process to the next will almost always have a bottleneck of sorts. The more complex the system and the more variation that occurs in the system, the more this bottleneck will impact the output of the system.

Let's look at an example. Below in Fig 3 the flow is in a business which has a software department. This department is separated into teams that perform a step of the process, rather than into cross functional teams.

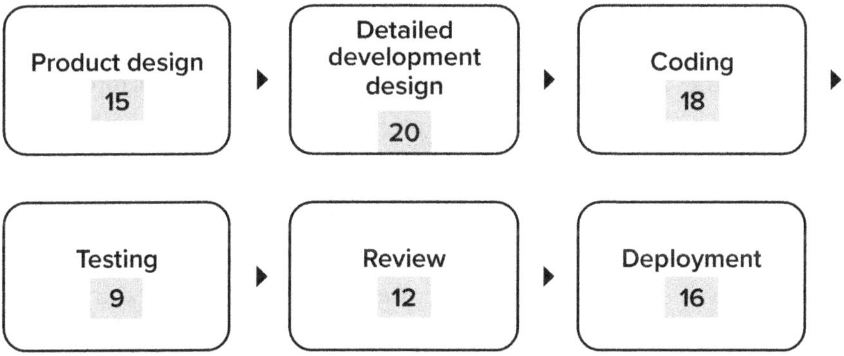

Fig 3

Assuming a piece of work, such as a new feature, has to go through all the teams, the numbers above show how many of those features each team can complete in an average month. Looking at the numbers of this department, there can be a maximum of nine new features per month. And you can be sure the work will be piling up in front of the testing team.

When teams are viewed individually rather than by the result of the total output, it's easy to miss the bottleneck. Poor data also doesn't help. Once the bottleneck has been identified, effort needs to be put into increasing the output of that team—in this case, the testing team. Once the teams have been optimized, the business should also look upstream and downstream. What are the other teams doing (or not doing) that is impinging on the testing team's performance? Once the other teams are supporting the testing team, only then should more investment in the testing team be considered, to get the most value out of what already exists before spending more. And of course, after all that effort, the testing team may now be performing better than one of the other teams, so the bottleneck should be reviewed again. If these activities alone could get testing team's output up to the same level as the review, that would be a 33% increase in output—33% more output for the entire department just by increasing the bottleneck's capacity.

This environment makes for a great opportunity to introduce a concept known as the Theory of Constraints. However, it wasn't developed with a modern software development team in mind, so let's take a look at how this concept is actually applicable here.

In any environment with a mixture of skillsets, there will be those who are the most experienced or the most productive people on the team (I'm sure if you've been there, you can picture them). Sometimes it can be hard to know exactly what the difference is; you might be aware that these people can achieve more than others, but it's hard to spot the difference unless there is a system with great visibility and you're really looking for it. Most software development teams will have a top person or two who are really productive, and then a few others who are decent performers and are productive but limited (because tests, reviews, difficult tasks, etc. have to go through them), and then a couple of fairly junior people who seem productive but whose actual output is low because they are limited by all the others.

The top devs will be fully loaded, or even overloaded. I don't know if there's a software team in the world where these people will only work 40 hours a week. They nearly always do more, and they are essentially gating the productivity of the team as a whole. In this situation, because the top few are the bottleneck, everyone else might slow down their work because they don't have much to work on. They might find other things to keep busy, or at least look busy, but busy doesn't mean productive.

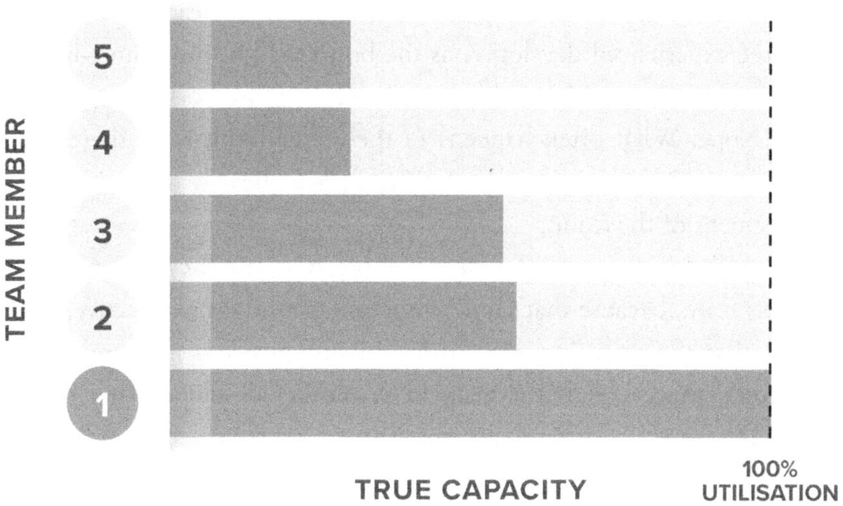

Fig 4 For simplicity, we will use a team of 5. 1 is the highly skilled dev mentioned above (and the bottleneck) 2 and 3 are the mid team, with 4 and 5 being the junior devs. The diagram shows the true capacity loading of these team members.

It can be difficult to see where the work is actually going with the junior members of the team. One of the worst situations is when juniors search out extra work that hasn't been planned into the iteration, since they want to remain productive. What they don't take into account is that, quite often, there are other aspects to these extra things they pick up. While the extra work might give the juniors something to do momentarily, it also needs to be designed or reviewed, which further loads up the bottleneck. The bottleneck people can only complete so much work, and the juniors are now wasting their time with things that are not necessarily valuable or relevant in the moment. Therefore, the bottleneck's capacity is full or overloaded, thus limiting the output of the entire team.

This is one of the major issues with the way that software development teams plan sprint work. They plan the sprint based on the capacity of each individual, as opposed to the potential output of the team. We've identified the experienced developers as the bottleneck of the team—but do we make sure to optimize their output, and reorganize others to support them? Nope. What often happens in these environments is the experienced devs are also the team leaders, or at least heavily involved in the management of the team.

Naturally, these overloaded devs will begin asking for more resources for their team, because that's how it appears from their perspective. They believe that the team is overloaded, and simply needs more resources. But would that help? If the team hires another developer, that person would then need to get caught up to speed with how the business or team operates, learn the software, perhaps learn how to write tests, etc. —they're not going to be particularly productive immediately. This new person's work would need to be planned and reviewed, and they'd need to be onboarded and trained—which nearly always falls on the team leaders and experienced devs. That additional work would either replace the work they could be doing (reducing their output and that of the team) or they end up working more and staying later, increasing the feeling of being overworked even more. So, a month or so down the line, even with an added staff member, the experienced developers who are the bottleneck are just going to feel like they need even more resources again, because that new person doesn't add any useful, productive capacity to the team.

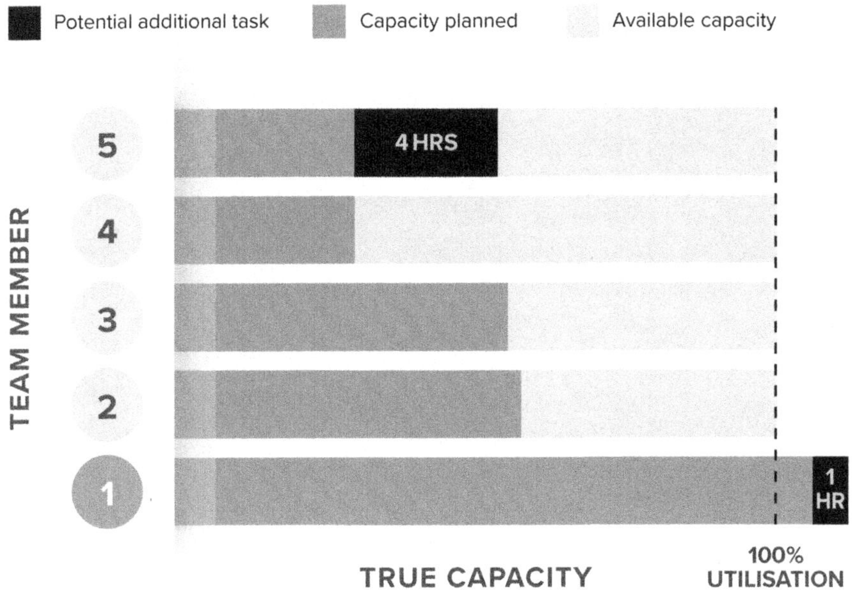

Fig 5 Other team members, such as 5 can't take on any more work without adding to the bottlenecks workload or replacing work already planned for the bottleneck.

CHAPTER 6

Bottleneck Rules

1. Ensure there is a steady supply of work for the bottleneck
2. Offload the bottleneck from unnecessary tasks
3. Plan others' work around the bottleneck
4. Ensure quality inputs into the bottleneck

There are four critical rules for effective management of the bottleneck in a software team or company. Bear in mind that the bottleneck could be anywhere in your process—it may be designs, or it could be testing, or perhaps integrations—this is going to depend on the nature of your software, whether it is a product, project based, or there to integrate with hardware. The first step will be to identify the bottleneck of the business. You will likely already have a hunch—it is most likely where work piles up, and the steps immediately afterwards often have plenty of capacity. Identifying the bottleneck and implementing the four rules should lean out the process.

The bottleneck in most software companies today, especially software product companies, is development; but merely declaring the bottleneck will do little to improve the output. You can still follow the four rules for development as a whole, but you will quickly find it apparent that the few expert developers (usually one per team) are the real bottlenecks. Because these teams are largely independent, their work doesn't often cross paths, making these experts the best place to focus, and that's exactly what we'll do for Xyphyr.

The rules can technically be followed in any order as there is no dependency (Rule 2 does not require completion of Rule 1); however, the listed sequence is the most effective order of implementation, being the easiest to get people on board with.

Rule 1: Ensure there is a steady supply of work for the bottleneck.

Before we try do anything for the team itself, we need to ensure that there is a consistent supply of work for the bottleneck. Given the backlog in most software teams, there generally won't be a shortage of work to be done, but all pre-requisite steps should be clear so that the bottleneck always has tasks to do. This may occur in the way sprints are planned or work is allocated; the worst situation is a team leader or project manager sitting on a task which should be made available to the team and the bottleneck. To ensure this availability of work, there should always be work ready for the bottleneck. How you define work that is 'at the ready' will depend on your team. If the bottleneck does the designs, then it might mean having enough user stories or customer requests to design. If the bottleneck does the coding but not the designs, it will mean ensuring there are enough designs finished and ready for dev work. This first rule is simple but critical. You may already be in this situation (congrats on your first step to bottleneck management!); however, if not, it is an important place to start.

Rule 2: Offload the bottleneck from unnecessary tasks.

So, what can be done? The first step is viewing the bottleneck's work with a critical eye. Depending on the people involved, this can go one of two ways: either the experienced devs say, "Oh, great!" and work with you to figure out what they can offload, or they resist giving their work to others for fear it will get mucked up. The most critical element in getting others to support the bottleneck is understanding that the efficiency of individuals doesn't actually matter—and this can be a difficult shift in perspective for some. The efficiency of the team as a whole is what really matters. Although it seems counterintuitive, it makes sense to take a task that might take the experienced dev two hours to complete, and pass it down the line to the mid team, where it might take them 6 hours to complete. As far as anyone can see, that feels incredibly unproductive; but although it takes much longer to do, it's actually increasing the team productivity.

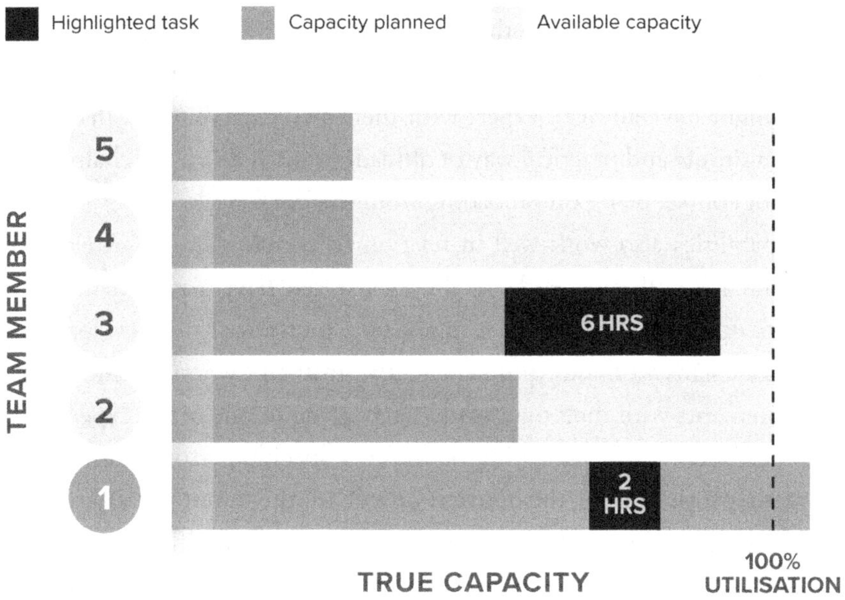

Fig 6 As work is offloaded it will likely take longer. This makes individual tasks less efficient, while making team output as a whole more efficient.

The team mindset needs to shift. The more pressure there is on delivering on time, the more quality issues will arise, and the less comfortable people will be with moving work away from the experienced developers—even though the teams with those issues are the teams that need it most. The ultimate goal here is to pass things down the team, from left to right in the diagram.

Capabilities

The first step in changing the way people see and manage capacity and task assignments is to assign tasks based on capabilities, and also to assign capabilities to certain people. A team will label what kind of work the experienced devs could do, what kind of work the mid team could do, and what kind of work the juniors could do, and allocate that to a capability. In some cases, you will have specialists; for example, the mid team might have an SQL expert with their own capability for that. This step is a simple and practical way of offloading work down the chain—the focus, of course, being on offloading from the bottleneck.

Capabilities also work well in relation to people who are external to the team who allocate work, as they will often have a person in mind to work on specific tasks. There might be a pretty well-run development team that plans work out quite efficiently, until an owner or senior manager interrupts with their own work, likely given to one of the experts. Or even customer service assigning tickets to a specific person because they know they'll get it done the quickest or will ask the fewest questions. This often happens when team members are mentioned specifically by name. "Oh, Mike is really good at that," or, "Emma's by far the most efficient at this work. Can she do this for me?" The moment people are named like that, things tend to fall more heavily on those resources. Using capabilities removes this pressure and allows a better way for tasks to be allocated and spread amongst the team.

Leadership

Now that we've addressed the dev work, let's focus on management and the team leaders. Some excellent developers lack the people skills to be excellent leaders. And the tendency that many people have to give leadership to the most experienced person can be bizarre and destructive. In many companies, someone will get promoted to management or leadership simply because they're the most technically capable person, regardless of how well they can actually lead others. So, someone who is not the best leader is now leading others—doesn't make a lot of sense. You've also now taken a whole pile of productive, top technical skill work off somebody and replaced it with other tasks that aren't really productive compared to what they've been doing.

In many teams, it seems that people in the mid team actually have better people and leadership skills, along with a desire to actually be placed in that role. Some of these mid team developers are actually great candidates for leaders. They, for whatever reason, are never going to be the world's best developers, but they have enough skills in all areas to be able to coordinate the team very well. If a team is already being run by the experts, shifting that leadership work down to one (or even a couple of) the mid team members, can massively increase the capacity of that team due to the freeing up of the bottleneck. Team output can easily be increased by 20% or more with that one simple change.

The other side of the leadership discussion—perhaps the more sacred cow to slay—is that 'whoever's in charge has to be paid the most.' This seems to be something that companies inherently believe is some kind of law of nature; that people can't possibly manage someone who's paid more than they are. This is a hangover from old-fashioned hierarchies, and just doesn't seem to go away. The truth is this: You can pay people as appropriate, taking into consideration market rates, and treat leadership as just another skill. Leading a team doesn't mean you automatically have to be paid more than someone who's more technically adept but has fewer people skills.

Overall, the key is identifying the bottleneck in the process, then making a concerted effort to plan workloads based on capabilities. Sit down, look at the tasks people are doing, and identify any activities that use the bottleneck's time—even those outside of dev work. Using a systematic approach, make a conscious decision to move work between people to best manage the operational work of the entire development team.

Rule 3: Plan others work around the bottleneck.

Our next step to maximise the value output of the bottleneck is to plan the team's work around the bottleneck. When planning a sprint, the bottleneck's work should be considered first. The goal is to offload everything we can from the bottleneck (Rule 2), then plan other work into the sprint which doesn't require the bottleneck. We should not fully load any other members of the team, either. There is always time lost in handovers between steps, whether they require some explanation or one of the team members isn't available, etc. For this reason we want our non-bottlenecks to have a bit of spare capacity to be able to be flexible around the bottleneck. Practically speaking, this time might be spent waiting on a handover that is coming soon. Let's say there is a dev who has their technical designs and tests reviewed by the bottleneck, and then that dev works on their code, which is then reviewed by the bottleneck. We want that dev to be ready and available when the bottleneck reviews their work—consistently and reliably—not when it's convenient for the dev. The last thing your bottleneck needs to hear is the dev dragging out, "Oh, yeah, I know we said 10:00 but I'm just finishing up this other thing, can we do it at 10:30?" . . . and then come 10:30, "Sorry, just give me 5 more minutes," . . . and then 17 minutes later, "Okay, now I'm ready." An hour of productive time from the bottleneck vanished because one person couldn't manage their time, or show integrity for promises. Disrespect to others' time like that is bad on the best day, but now it's limiting the output of an entire team! It would actually be best for the dev to not start any new work and muck around in their email or on their phone for half an hour before the review meeting with the bottleneck, to ensure they're on time. The thought of that will make any accountant, project manager, or busybody reading this very upset, but being upset doesn't mean they're right. Any time you question this, imagine this extreme scenario in the emergency services: No one wants a fire engine to leave the station 17 minutes late because firefighter Larry just wanted to take 5 minutes to finish testing the equipment on the other truck.

The work that must be done by the bottleneck should be planned first, followed by planning the work for the rest of the team. When planning the bottleneck's work, we can also take into account the type of work that needs to be done. We will look more at managing that later on.

The Professor's Jar

There's an old story, with no known origin, about a professor and a jar. This version by Dr. Stephen R. Covey is a great rendition:

> One day, an expert was speaking to a group of business students and, to drive home a point, used an illustration I'm sure those students will never forget. After I share it with you, you'll never forget it either.
>
> As this man stood in front of the group of high-powered over-achievers he said, "Okay, time for a quiz." Then he pulled out a one-gallon, wide-mouthed Mason jar and set it on a table in front of him. Then he produced about a dozen fist-sized rocks and carefully placed them, one at a time, into the jar.
>
> When the jar was filled to the top and no more rocks would fit inside, he asked, "Is this jar full?" Everyone in the class said, "Yes." Then he said, "Really?" He reached under the table and pulled out a bucket of gravel. Then he dumped some gravel in and shook the jar, causing pieces of gravel to work themselves down into the spaces between the big rocks.
>
> Then he smiled and asked the group once more, "Is the jar full?" By this time the class was onto him. "Probably not," one of them answered. "Good!" he replied. And he reached under the table and brought out a bucket of sand. He started dumping the sand in and it went into all the spaces left between the rocks and

> the gravel. Once more he asked the question, "Is this jar full?"
>
> "No!" the class shouted. Once again he said, "Good!" Then he grabbed a pitcher of water and began to pour it in until the jar was filled to the brim. Then he looked up at the class and asked, "What is the point of this illustration?"
>
> One eager beaver raised his hand and said, "The point is, no matter how full your schedule is, if you try really hard, you can always fit some more things into it!"
>
> "No," the speaker replied, "that's not the point. The truth this illustration teaches us is: If you don't put the big rocks in first, you'll never get them in at all."

This analogy illustrates the importance of planning bottleneck tasks alongside the tasks of the rest of the team. The biggest tasks or projects need to be planned first—do they need to go through the bottleneck, or are they not that critical? Once the bottleneck's time is planned, it's like the jar full of rocks. The next step is figuring out what the mid team can accomplish, which is like the gravel. And finally, the juniors fill in the work like the sand and water. Unlike the professor's jar, it's never going to fill up perfectly, nor do we want it full as discussed earlier. It's impossible to get that kind of full balance on a development team, where everyone is fully loaded to capacity but not overloaded.

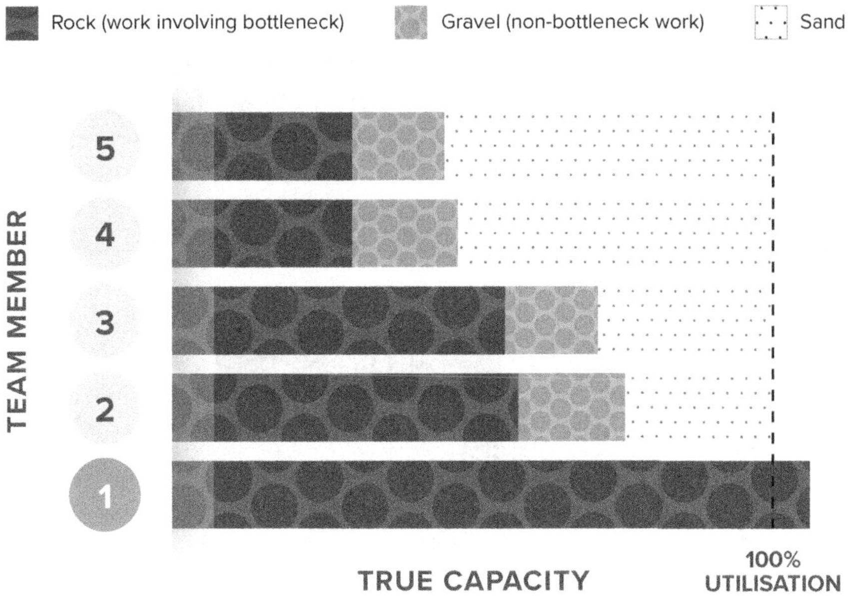

Fig 7

Rule 4: Ensure quality inputs into the bottleneck.

Quality Assurance

Once you've taken whatever planned work possible from the bottlenecks and passed it to other team members, the next step is to ensure that the work they do (which interacts with the bottleneck) is the best quality possible. This can be as simple as getting the prerequisites right—ensuring that anything that is about to be handed to the bottleneck gets an extra review, to make sure everything is present and correct. It's easy to skip scoping or design details when under pressure; if you don't spend as much time scoping, the experts will have to re-do half of it once the work passes into their hands. Having the most valuable resources burn valuable time because the poorly detailed design given to the juniors meant they didn't code the right things is an incredibly bad trade of value for time.

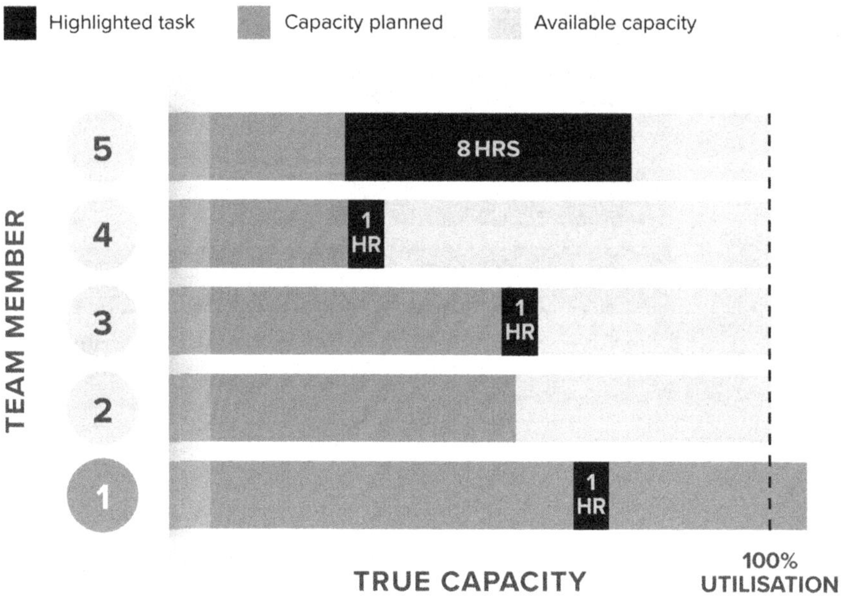

Fig 8 The review steps are shared amongst the team to reduce the loading on more expert resources.

Another important step is adding in extra quality steps. Passing lateral quality checks at each step will make the most of this step. For example, if a junior coded something, another junior should complete the first review. Then it can get passed up to the mid team, and from there vertically again to the bottleneck. This means that most of the basics should theoretically be sorted out by the time the work gets to the bottleneck, so the bottleneck will really only need to review the expert bits. This also gives the juniors and the mid team the chance to upskill and deliver higher value into the future. If you manage this, the notes made by the experts should also be read by anyone else who reviewed the task down the line, so they can also gain valuable insights from the expert review.

Enforcing Bottleneck Operation

Now that you've made the rules of the bottleneck operation, who will enforce them? In most cases, this role would be the equivalent of the person planning the iterations, whether sprints or other ways of managing team work. The role can vary; some like to have one person enforce the operation across multiple teams. Others like to have operation of each team enforced by one of their own team members. In this case, someone would take responsibility for tracking and managing the work, along with the way it gets planned within the team, which tends to be the more common method.

Case Study: NHS

The National Health Service (NHS) of the UK is a great example of use of the Theory of Constraints. The NHS uses bottlenecks as a primary way to improve its systems and encourage its components of departments, hospitals, clinics, and others to use these methods to improve their practices, processes, and procedures. The NHS advice about this is publicly available on its training resource website. The NHS is confident in the impact of bottleneck management being applied to healthcare, and has stated its effectiveness in reducing waste, decreasing waiting times, and increasing the throughput rate of patients.

The NHS is somewhat infamous for being largely underfunded considering the UK is a wealthy country; however, despite this reputation its performance is still average amongst OECD countries. As previously mentioned, everything possible should be done to gain maximum performance from the bottleneck before investing more into it—something that would be critical to a government department that can't lead initiatives with the approach of throwing money at it. I speculate this funding limitation is what led the NHS to bottleneck management as an approach to improve output without having to invest more.

Amongst the advice from the NHS on improving components using bottleneck management is looking at the flow of patients through the hospital or clinic and identifying the constraining factor. One such case of this is the Radcliffe Infirmary in Oxford. The hospital implemented bottleneck management to increase the throughput of its operation, essentially to get a better result for not just the patients but everyone involved—the staff were overworked to a dangerous point. Elective patients and their operations were getting cancelled because there wasn't space available. At the time, Radcliffe was one of the lowest performing hospitals in the region.

Identifying the bottleneck was an interesting part of the process with the internal team initially identifying the bottleneck based on their experience as either neurosurgeons or anaesthetists, with both parties believing it was themselves. This was a reasonable place to start, as there have been clinics where this was the case. However, it can be dangerous to jump to conclusions or assume the identity of the bottleneck.

Once Radcliffe's team went through the proper processes of studying the flow of patients through the hospital, it turned out that the constraint was, in fact, beds. Given both the facts that the NHS is low on funding and adding more 'beds' isn't as simple as buying more actual beds, the required renovations and expansion of the hospital wouldn't be an option.

Instead, the team began with the ward staff in the stressful position of managing the use of beds. The current management of beds was chaotic and led to many cancelled appointments and wasted bed usage. To combat this, they assigned a bed scheduler—a person whose sole job was to ensure effective bed scheduling—to look across all beds and patients, and ensure the bed capacity was maximized. The impact of the bed scheduler was proven when the role became temporarily vacant and the old symptoms returned almost immediately.

This is not the only case across the NHS where beds are the constraint—in fact, it's so common that the NHS provides specific advice in the online resources about how to decrease the time from operation to discharge by increasing rehab and its effectiveness, as well as other steps. These steps help patients recover faster so they can be discharged from the hospital or clinic as fast as possible, which then allows more people to use the beds.

Outcomes for the Radcliffe Infirmary, leading them to become one of the top five performing hospitals in the region:

- Cancellations dropped from 64 in the previous year to none.
- Out-of-operations surgeries were reduced, decreasing surgeon stress and associated dangers.
- Operating theatre utilization was increased from 73% to 88%.
- Number of patient operations increased by 15%, resulting in 13 more elective surgeries per week.

CHAPTER 7

Introduction to Variation Management

Planning work to optimize the bottleneck is the first step of increasing the pace and quality of work with the same resources. But that step alone won't lead to success because, "No battle plan ever survives contact with the enemy" (Helmuth von Moltke, Prussian General). The reason those battle plans don't survive is variation, of which there are two major causes.

1. One is the uncontrollable external party. In the context of the quote, this is the enemy; in our context it's mostly users, customers, or the market. We limit the impact of this by committing to a specific piece of work as late as possible (hence sprints, and ideally shorter sprints), but we'll get into that more in the next section.
2. The other is natural variation that occurs in any environment. "Anything that can go wrong will go wrong," Murphy's Law, if you will.

There are management practices all over the world designed to manage these in one way or another—an example being 'Lean in the Toyota Way,' where the goal is to reduce all sources of variation. They start with the biggest variation and work their way down.

There is an analogy of a pond representing inventory, and rocks representing sources of variation. For smooth sailing in a rocky pond, you need a lot of water, but that is expensive. So you lower the water level and remove the large rocks that present themselves. Once you have done this, you can lower the water level further and remove the next size of rocks, and so on.

As you're surely aware, this method has worked very well for Toyota, and other production line manufacturers. The two problems with this method in Software Dev are:

1. Firstly, a production line has far more known variables due to its repetitive nature. The closest software parallel is system setups if they are not automated, where a technician must configure the system for a new customer or user, and even then, it is far from a perfect parallel.
2. Secondly, it was a hugely expensive undertaking for Toyota to implement this system, one that is unfeasible for most companies.

Where does that leave us? We need a way to protect our team's—and therefore our bottleneck's—work from this natural variation. It would make sense to add a buffer to task estimates so we have some room for variation. You're probably thinking, *Yeah, my team does that, but it doesn't do enough to really manage the variation because we still feel the problems of it.* The reason is, adding buffer to individual tasks is a bit like the individual efficiencies discussed earlier: they only benefit the individual and are rarely passed on to the team. If people have longer to do a task, they tend to slow down when everything is going according to plan; even when they finish early they are often not in a hurry to hand it over. Why should they? There's no urgency when everything's finished ahead of time! This way, the gains of going quickly are lost, but the losses experienced when some tasks run over accumulate—meaning we only ever get behind, rarely ahead.

In addition, if people can't see the collective loss of the team in a simple and visible way, they can't respond to it. Skilled Scrum Masters and experienced team leaders will have built-in intuition about the progression of the sprint which will allow them to respond, but the rest of the team will likely not share the enthusiasm; and even when they can respond, do they know how to? So we need to add a buffer, which is visible to all when it is being eaten into, and only by the planned tasks (not new tasks being added mid-sprint) when the buffer is running low. We need a planned response that carries different actions. Enter Variation Management.

Note: Variation Management is also a critical element of Pace; however, it can be applied to sprints too. We'll apply the rules to sprints to make a relatable point, and to give the option of staggering the Pace install phases (see the implementation section of this book).

CHAPTER 8

Variation Management Rules

1. Ensure the bottleneck's work is buffered
2. Make the passing of time visible
3. Use an alternate response system
4. Buffer Logging

Rule 1: Ensure the bottleneck's work is buffered.

The first step in Variation Management is to establish the buffer. A buffer is a section of planned time between the end of the estimated time in a sprint and the actual end of the sprint. This time is there to protect against variation. In order to keep with the tradition of fitting work arbitrarily into week-based sprints, I recommend you set the sprint end to be the same as your current sprint (for example, 2 weeks). In this case, you will need to reduce the amount of estimated work in the sprint.

How long to set the buffer? Yet again, 'it depends' serves here. If you are in a highly variable environment (perhaps dealing with new, cutting-edge development, or interrupted with a lot of customer feedback) you will want a longer buffer than a less variable environment (perhaps customisations that are similar but different to customisations done for other customers). A good rule of thumb is to start with 30% of the planned time, which can then be adjusted up or down based on how much of the buffer you use each sprint, or if you go over the buffer! Remember to base these decisions on usual scenarios rather than the unusual—don't use the example of one time 3 years ago when something took 5 weeks rather than 2 days.

Using our same diagram to visualise this, imagine that—rather than average capacity—the bars now represent planned work.

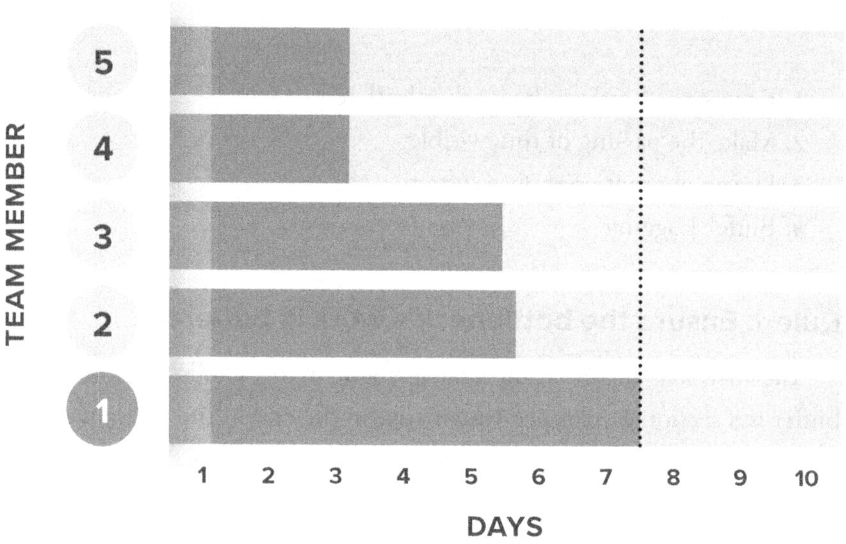

Fig 9

In this case, the last 3 days of the sprint act as the buffer. While it might feel like we will get less done in the team because of the 3-day buffer, people naturally build in time to their personal task estimates to account for variation—otherwise those estimates were entirely unrealistic to start with. The reason the individual task buffers aren't effective here is because people will use the time they have available to them. If you think a task will take 4 hours, so you estimate 6 to give yourself a buffer, when you come to do this task, you will think, 'I have 6 hours to do this', not 4. This way the buffer disappears. We will have to be strict with people to make sure they take their task buffers out of their individual tasks. If you have a team buffer and people are still building in individual buffers, either they are reacting to the buffer still being too short, or the management is measuring them on individual efficiencies (see: busy-ness). If it's the former, adjust the buffer; if it's the latter, adjust your managers.

Once the buffer is established, you can be far more confident that the output of the bottleneck will be protected and reliable in most cases. You can also fill up the other resources with gravel (non- bottleneck) work to the new shortened length, as the buffer will give them the time to be flexible around the bottleneck.

If you're obsessed with busyness and haven't got over that yet (you will by the time your team is running Pace), you can have the introduction of sand tasks, which are highly interruptible tasks which don't involve others—maybe writing an article for marketing to put in the blog. These can be done by non-bottlenecks when they have free time allowed by the buffer.

There will, of course, be cases where the existing buffer is not enough, and we will need to respond to its condition. This also lets us shorten the buffer if we run the next rule effectively.

Rule 2: Make the passing of time visible.

The intuition of those running or managing the execution of the sprint mentioned earlier is best summed up as being able to compare real progress to planned progress, and make adjustments for it. This is similar to skilled project managers being able to tell whether things are on track, and even knowing when the report is lying (because people are intentionally trying to buy time, or unintentionally not taking proper consideration of the state of their tasks). The other part of this is learning how to properly react to this, covered in Rule 3: Use an alternate response system.

We want everyone to be able to see the passage of time against the progress they've made, so that the whole team is informed and on the same page with decisions and updates. Adding visibility also drives more useful communication—if people already have the state of their task visible, any communication (for example, at a stand-up) will be around what to do about the task, if anything. Whereas without the visibility, most of the communication is focused on merely updating what has been done—which is a waste of time, not useful for team progress management, and boring for everyone else (meaning a lot of zoning out and 'waiting to say my bit').

To make the passage of time visible, we need to be comparing the progress made to the estimated progress. However, as we know, the bottleneck gates the output of the team as a whole. They are the only fully loaded resource on the team, which means their actual progress is what really matters, not the estimate.

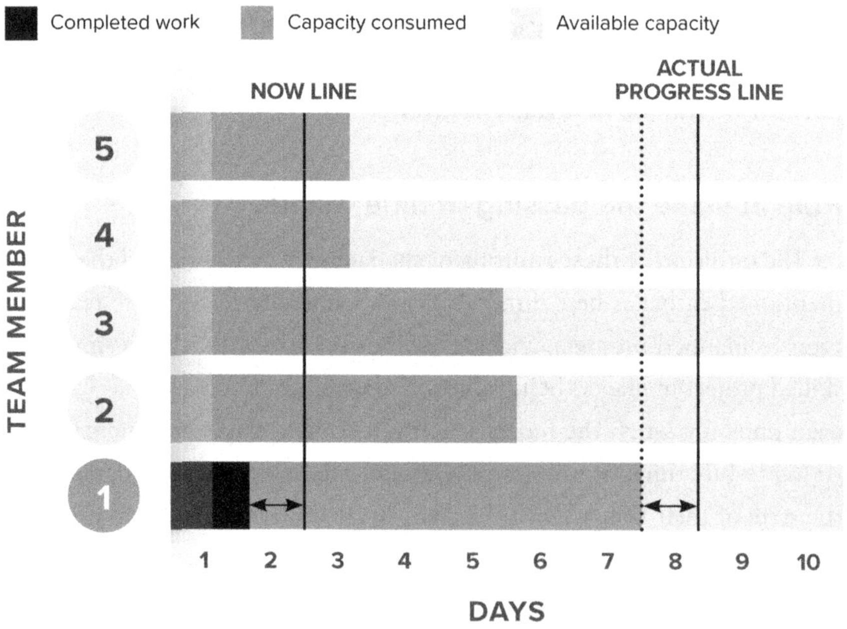

Fig 10 The difference between the Now Line and the actual progress shows how much buffer is being used.

Imagine a planned work chart (either a physical one on a whiteboard or pinboard); each day when you have the stand-up, the bottleneck marks their actual progress. Then, to represent the passing of time, hold a ruler or attach a piece of string to the top of the board where 'today' appears in the sprint, called the 'Now Line'. The difference shown is actual vs. estimate. Some may wish to calculate this—and when the process is run in software, that's done for you—however, provided the sprint chart is to scale, you can use a ruler. The difference represents how much of the buffer has been used, so you can move the buffer indicator (pin, string, etc.) to give your team a visual of how much of the buffer is used.

Keep track of things by comparing the buffer usage to the progression of the sprint. If it's day three of the sprint and two out of your three buffer days are already used, you will have to make changes to get back on track. With the buffer visibility making things clear to everyone, the stand-up meetings can focus on how to manage the risk of the whole buffer being used, and therefore the sprint being unsuccessful and taking longer than the allotted time. It also helps team members internalise the Rules for Bottleneck Management by making clear the extent their interactions with the bottleneck matter.

The team is now focused on completing sprints as a team, rather than as individuals.

Rule 3: Use an alternate response system.

There are project management techniques which help projects get completed on time and on target. Some of these techniques, like the visibility of progress discussed in Rule 2 above, are easy to implement on a day-to-day basis, and people generally like them (if in doubt of any measure, make it visible—people love simple data, especially high performers). However, there are also project management techniques which people don't like so much because they are more annoying to the individual.

[Note: Deadlines don't count here as a technique; rather, they're a hammer when project managers can't get anything else to work.]

These disliked techniques are often deployed in high-pressure environments, but don't tend to go over well with relatively autonomous teams and individuals in software teams. So, how do we reap the benefits, without the shortfalls? One major shortfall of these techniques is that they work for a while, but then people get sick of them and they are no longer effective.

This set of alternate responses include things like: updating estimates as you work, informing team members when to expect handovers, pausing less important tasks to subordinate to more critical tasks, etc. The techniques themselves will vary from business to business and from team to team, depending on the nature of the work. The commonality is that we want people to turn to these actions only when they need to—when everything is ticking along fine, it's not necessary. The problem with this approach is that, in most cases, it's hard to tell when things are ticking along fine and when they're going off track, so we often react too late, once the sprint has already gone off the deep end and no reasonable effort will save it. Depending on the criticality of the sprint and the culture of the team/business, this means either unreasonable effort (goodbye nights/weekends) or no effort at all because, 'oh well, what can we do?'

With a buffer-managed bottleneck, for which we are actively tracking the passing of time, we can now employ this alternative second set of actions proactively—before all is lost! The major difference for any team running this alternate response is that they will reap the benefits of a world-class project manager without hiring one (and their sharp stick).

The first step is deciding what actions the alternate response will include. This is best done with the team members themselves, as they have to be willing to execute what they commit to; once the alternate response is in place, 'oh, I'll just focus on that one today,' or, 'I'll work harder on it' are no longer accepted as responses to high risk tasks.

The second step is to decide when to employ the alternate response. The only way to do this effectively is with a set point, which doesn't change based on people's opinion—the ability to talk our way out of doing something we should is amazing and destructive. You can structure this around the visibility of the buffer and how much of it has been used. We cut our buffer into thirds—which we name Zone 3, Zone 2, and Zone 1—which work like a Red-Amber-Green system. Zone 0 is the infamous Zone where there is no buffer left. Depending on the severity of a sprint's deadline, some teams will employ a second set of alternate responses at Zone 0, which are intended to be very harsh. Most teams won't need this, and the second set of alternate responses will suffice. Zone 1 is where the alternate response comes in; once the buffer is in Zone 1, the questions at the stand-up (and actions that follow) change to reflect the Zone 1 risk. For example, "Is there anyone on track who can help accelerate this task?" It is critical to remember here that, while we are running the buffer on the bottleneck's work, they don't work in isolation—therefore, others on the team will impact their zone state by delaying them with things like handovers. Once the buffer arrives in Zone 1, those previous suggestions about how team members should treat bottleneck tasks are no longer suggestions . . . they're required!

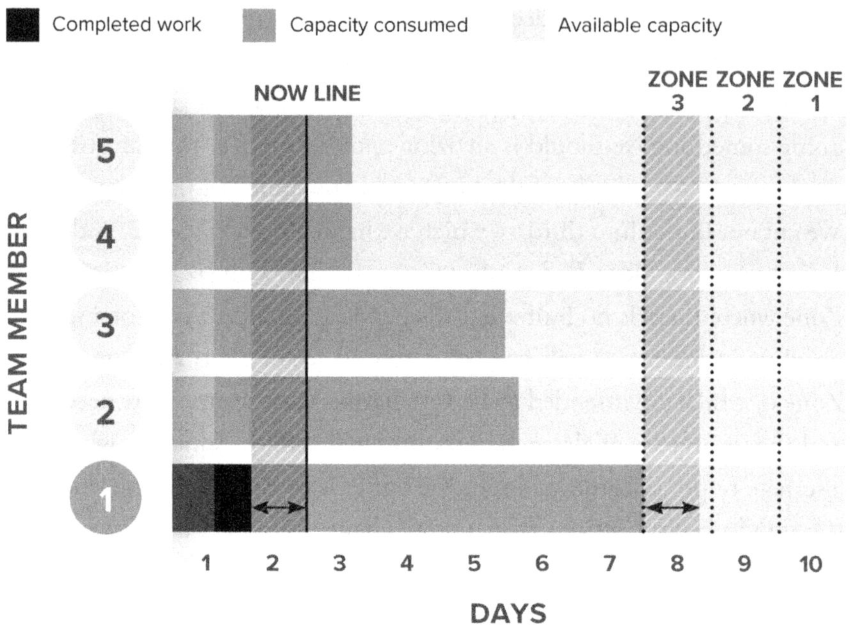

Fig 11 The team is in the same state as Fig 10, however, now has Zones to add further meaning to the buffer use.

Although Zone 1 triggers the alternate response, Zones 2 and 3 are also important. They add to the visibility of the passing of time. Once people have experienced a few Zone 1 hits, they'll realise that Zone 1 isn't fun. Keeping out of Zone 1 quickly becomes an informal team goal. The boundary between Zones 2 and 3 give team members a warning on the way to Zone 1, and let them know they should do what they can to get the tasks back on track informally before they're required to make a large change.

The alternate response system can be adjusted from time to time to reflect what works for the team and the individuals in it. However, these alternate system actions should not be adjusted immediately after a Zone 1 hit; in that case, it's likely that the recent effects felt would taint the rule adjustments accordingly. If the Zone 1 actions aren't working to get tasks back on track, you will need to run some informal diagnostics:

- Are people actually following them? (People should be able to be observed following the planned rule behaviours when in Zone 1.)
- Are the actions ineffective? (When people are doing the actions, there doesn't seem to be any effect.)
- Are the estimates too low/buffer too short? (People are doing the actions and they work, but it is always too little too late.)

Once you have a grasp on how to respond to the risk of an individual sprint going off track, you can use the visibility of buffer use to make long-term decisions to improve sprints, which brings us to our next Rule.

Rule 4: Buffer Logging.

There's no point in making the same mistakes over and over again, just as there is no point in letting the same sources of variation interrupt your sprints over and over again. It can be difficult to pinpoint exactly what it was that de-railed a sprint, and even more difficult to know exactly where the task blew out that wrecked the whole thing. The previous three Rules give accurate live information about when a sprint is going off track; for example, going from Zone 3 to Zone 1 in two days is a red flag!

It's worthwhile to review the information, and see what happened to the sprint—in this and all cases. Reviewing and analysing the information over multiple sprints can limit the impact of sources of variation on the team. Let's say it turns out there's a consultant or customer service rep who is frequently giving incomplete or incorrect info to the team. We might feel like we know who the culprits are, but having data saying 50% of Zone 1 hits are a result of that incorrect information, we can be sure it's worth training this person or updating their processes. It's a pity to have to say, but in some teams, data like these may be useful or required to back up arguments, depending on the state of office politics between the teams and managers.

Buffer logging is as simple as keeping a sheet beside the sprint board, or an Excel sheet to track the cause and frequency of Zone 1 hits. This is useful information for the team during sprint reviews to discuss what can be done internally about the causes, and then escalate them as necessary to causes which lie outside the team. We aren't always sure what will work when discussing solutions in theory, either. Instead, you can test actions, by noting them beside the Zone 1 reasons. The next time the team is heading for Zone 1, they should take a look at the common list of hit reasons, try to identify the cause, and apply the test action. If it works, you can upgrade it to the list of alternate rules to be required in Zone 1, and used at the team's discretion outside Zone 1. This builds a process of ongoing improvement owned by the team, which doesn't require a huge amount of time to maintain and update.

Case Study: Lufthansa

Critical Chain Project Management (CCPM) is an approach that incorporates bottlenecks in both project planning and execution. In many cases, this is not relevant to software since the projects they are delivering are either one-off Waterfall style projects (which if applied to software would lose the iteration benefits from Scrum) or they are the other extreme—large but fixed manufacturing projects with known steps completed over and over. However, in the Aircraft Maintenance, Repair, and Overhaul (MRO) industry, CCPM is perhaps surprisingly comparable to software. This is due to the flow and task types being the same in each project (we would call iteration), but the unknown variability being extremely high. Let's look at an example and then figure out how we can maintain the principles while scaling down the project length into smaller iterations.

Lufthansa Technik

Lufthansa Technik is the MRO subsidiary of the Lufthansa Group, and completes what are known as 'checks' on aircraft for Lufthansa, along with aircraft for many other airlines. The primary types of checks are known as 'A' and 'C' checks. 'A' checks are more frequent, smaller checks—similar to receiving an annual service at the car dealer. 'C' checks, however, are significantly more involved and require pulling apart the aircraft and engines to go over them with a fine-toothed comb. In any check, but especially a 'C' check, the variability of what needs to be done is massive, and the project teams have no idea at first how long or complicated the check on each aircraft might be. This scenario has similarities to debugging in software, although the aircraft checks are preventative maintenance rather than fixing an aircraft that's not working. During the checks, an engineer often examines one part of the aircraft, only to find something else, and then something else beyond that. In addition, the aviation market is growing so rapidly that the new aircraft types entering the market are constantly requiring different types of checks and attention.

This creates a daily headache for many MRO project managers, and leads to an industry norm of firefighting. Going into a project, most MRO teams find it hard to know where the issues will arise, and what bottleneck will slow down each aircraft in particular. This leads to multi-tasking among the teams, jumping around various tasks, and worse, wasting the use of highly skilled staff. Estimating work time becomes a wild guess as any hope of accurate handover or completion times is destroyed by the Grey Time waste from handovers, multitasking, and waiting periods.

So, what are the rules[3] Lufthansa Technik uses from CCPM to create its "Magic Formula" as they call it, which freed it from all that chaos?

3. These are their rules that while different to the Pace rules, serve the same purpose in MRO.

Low work in progress.

Work is broken down into smaller packages, with fewer tasks in each work section, contrasting with the traditional major milestones. Breaking down the work into smaller packages allows the team to attain better visibility of progress and much faster task handovers. Once this work is broken down, the emphasis is placed on quick handovers to the next team, and quick starts on work that has been handed over. In any environment with deadlines or major project points, people tend to put off completing the work until they need to (known as procrastination or student syndrome). The way Lufthansa Technik combats this is two-fold:

1. First, team members are ready to receive their work and begin immediately. If someone in a chain of tasks finishes their task early but the next person is not ready to receive the handover or start the next step, the time gain is lost. If the first person is late and hands the task over late, there is also clearly loss on the project. This is true across every step, meaning that gains are always lost, but losses always accumulate.
2. Second, when planning and estimating work, the buffers added to each individual task are taken out and put at the end of the project. This relates to the first step and whether team members are making the most of the time with slick handovers. The total buffer of time can then be used to visualize and manage the project, which is the next major rule.

Variation Management.

The buffer is used for tasks which need it, allowing for things to take longer than expected and even for the odd blowout without pushing the project beyond its overall time plan. Project managers then monitor the status of the buffer and respond to a buffer with increasing risk, so they don't need to watch every task's progress, but only those that are consuming the buffer. The team can then give its own attention to resolve the difficulties at these steps and act proactively as tasks increase in risk, rather than reactively after the project has gone off the rails.

This new reporting and focusing method gives new attention to management reports, and productivity focus shifts away from ensuring high levels of individual productivity (which is the norm in MROs) to focusing on the lead time instead. The time an individual takes is irrelevant; the time the whole team takes to complete the project is what matters. All this visibility allows both the internal team and the external customer to have a clear and accurate view of the progress of the project so everyone is kept updated.

Don't start what can't be finished.

The third of the major rules is to ensure all prerequisites are cleared before any work is started. This is implemented by breaking projects into two stages. In the first stage, called Full Kit 1, the team depanels the plane for inspection, and completes a thorough evaluation of the state of the airframe and engines. Routine tasks are then carried out. Then, the team preps for the second stage by preparing Full Kit 2, based on actual findings for that individual aircraft.

Traditionally, MROs would rip into any non-routine work, following the age-old principle of "the sooner you start, the sooner you finish." But this only increases work in progress, and multitasking when the materials and necessary tools aren't available. The two stages to the project allow the team to ensure all prerequisites are cleared, so each task can be completed in quick succession and handed over smoothly to the next engineer.

CHAPTER 9

Introduction to Flow Management

As discussed in the previous sections, shorter sprints contain great benefits, but also create larger downsides in other areas. Essentially, the shorter the sprint is, the more "Agilely" the team is operating. This means that the team will experience both the positive and negative extremes of Agile.

Analysis of Short Sprints

Consider the extreme of a single-day sprint, as an exercise to make a point about the impact of both the positives and negatives. As the length of the sprint is reduced, the accuracy of time estimates is increased, because the length of time is shorter. This short sprint doesn't require sequencing issues of estimating future tasks for days or weeks ahead, contingent on the previous tasks as well. With the shortened sprint length, quality steps such as reviews can be introduced more frequently. When quality steps are delayed or batched together, they are less effective for two main reasons:

1. People are more attached to the work they have done when it takes more time or effort. Thus, they are less likely to make changes that increase the quality of the code, whether effecting the creation of bugs, or contributing to technical debt.
2. Reviews will be less thorough as the reviewer will tend to rush through the process, just wanting to get it done as the review length increases.

Benefits

Smaller and more frequent quality steps require less rework, work is more likely to be on the right track, and any rework is likely to be smaller and more contained. As the length of the sprint decreases, interruptions also decrease. People are less likely to feel the need to interrupt the sprint if they can just add their task to tomorrow's sprint. The longer the sprint is, the more likely it is that a pushy sales manager or founder who is excited to help out a loyal customer will want to get the work done immediately, by shoving the task in to a live sprint. Shorter sprints also allow for deployments into the code more frequently and consistently, which is extremely valuable for some. Sprints create much smaller, more frequent deadlines for work, when compared with traditional project management. More distant deadlines have the potential to create "surges," when developers have more time to let their work get away from them, creating a last-minute scramble as the deadline approaches. Ultimately all these steps are to limit the ability for external variation to impact the work we have committed to start, which reduces interruptions, task switching, pick up / put down, and rework. To paraphrase Moltke's quote, if no plan survives contact with external variation, we should limit the number of things planned, to limit the contacts.

Costs

With all of this considered, Agile will also have negative impacts when used to the extreme. In fact, the negative impacts can be extreme, because they actually outweigh all of the positives we've already explored when it comes to short-term sprints. I say this because a sprint less than five business days is unheard of, and most software companies use a standard two-week sprint.

Planning becomes far more difficult for managers because the plan essentially changes every day, with backlogs being chipped away at, day by day. This gives managers (and members of the other businesses' departments relying on development) very little visibility of what will be coming up in the future, even if that future is only a few days away. They have no way of knowing if the bug will be fixed by tomorrow, within the service agreement. Development teams already tend to be adverse to planning, so daily planning requirements might face significant resistance. In addition, this type of standup or planning would require all devs to work simultaneously, which would destroy the flexible culture which many software businesses promote and enjoy. Worse than that, everyone knows how "brief" meetings can explode to become anything but brief, and planning sessions can easily take up a very large chunk of the day.

It's important to view the way that bottlenecks would affect others during very short sprints as well. During very short sprints, it's likely that the people other than the person actively working on the task will be sitting around, waiting for their turn. Although those people could be loaded up with more tasks, the bottlenecks are truly the limiting factor when it comes to what can be done in a day-long sprint. This leaves the other team members with very little to do; they can't begin work on "tomorrow's" sprint, because that will just put the constraints further and further behind.

Finally, the short time frames with a daily deadline allow for little to no flexibility in timeframes for when estimates go to plan. This means that team members will either finish the work early and have nothing left to do for the day, or they will come up short and late on the estimate and feel pressure to skip other quality steps.

To conclude, short sprints bring huge benefits at even greater costs. But what if there was a way to get those huge benefits without the costs?

CHAPTER 10

Flow Management Rules

1. Small chunks of work
2. Use time buffers not deadlines
3. Limit work in progress
4. Have a relevant mix of tasks

Many of the issues brought about from sprints come from the fact that work is batched together into a chunk (normally one or two weeks' worth), which then has a deadline to be met. Essentially, this uses traditional project management, but in shorter batches or iterations to reap the benefits of such. But it's the hangovers from this traditional method that cause the issues.

Just the name itself—sprint—hints at a strenuous burst of activity (one that can't be maintained) toward a short-term goal. Then we stop, take a breather, and do it again.

More appropriate is a smooth and steady pace which is sustainable, without deadlines (which are largely arbitrary anyway, and only in place to drive time management). Who decides a feature set takes 40 or 80 hours for a team to produce, just because that's how many hours we decide to work in a day or week?

Why is there such a focus on deadlines anyway? Deadlines contribute to two productivity-killers built into human nature:

1. "Parkinson's Law"—the adage that "work expands so as to fill the time available for its completion." The longer we have, the longer we take to do something—whether we slow down or we take more time to polish, they're well observed behavioural patterns.

2. "Student Syndrome"—people start tasks as late as possible, eliminating any of the positive benefits of buffer built into planning and estimates, and putting themselves under pressure and stress.

What if, instead of this, multiple iterations (sprints) could be operating at any one time; as one finishes, the team rolls onto the next one. This allows the planning of the length of the iteration to be appropriate to the workload for that iteration. This would address the previously mentioned issue of people not having work to do at any given time during a sprint. But how would deadlines work with multiple iterations running simultaneously? They would become rather difficult to manage in this system.

So why have deadlines at all? The purpose is to help ourselves as individuals, and help managers to manage progress against something. With no time management, everything would take a long time. In this system, deadlines could be replaced with estimated time plus a 'time buffer.' Adding a time buffer and tracking the progress of time since the iteration began gives an appropriate bracket in which the iteration will be completed. A bracket is more sensible than a hard date. Given the multiple iterations being worked on by the team, some will take longer than estimated, and some will take less time than estimated, as is normal variation experienced in software development.

Over the multiple iterations, these variations will average out to be a smooth and consistent rate of development. This pace of development is then used in place of sprints for planning and delivery estimates. With no due dates to plan iterations around, work is planned against the capacity of the team. This helps determine if more iterations should be introduced to the time or if there is sufficient work for all members. For example, if there is a five-day buffer of work to complete iterations and there is a person or capability in the team that already has 35 hours' worth of work, any new iteration could only be given to the team if there are five hours of work or less in the new iteration for that person or capability. This system maintains an appropriate number of tasks to keep the team productive; however, not so many to allow work in progress to build or multitasking to occur.

The process of planning the iterations, estimating, and assigning the tasks becomes decoupled from starting on the iterations. Unlike a sprint, teams no longer plan the iteration and then immediately begin working on it. This way, planning iterations can take place whenever and however best suits the team. This planning process might be weekly, individually (when a dev gets, or needs, a break), or coordinated by someone in a specific role that plans iterations and queues them in front of the team, ready to go as the capacity become available.

In addition to planning, many other rituals that teams have surrounding sprints and Scrum also take place in Pace. Pace implements daily standups to discuss the status of tasks, primarily around the ones at risk of running over, discussing what is being done to escalate them or if there is another member of the team who can help. In some cases, this is just cleaning the task to be ready for the handover, minimizing downtime in the iterations and ensuring the task is finished as quickly as possible. Handovers should be performed like a relay race: the next person should know the handover is coming, and have everything they need to receive it and run with it.

A team working at Pace has a steady stream of work being completed and an equivalent amount of work being approved to be in play for the team. Anything that would once be an interruption can now be at the front of the queue to be the next iteration for the team to work on. A focus on 'chunking' down smaller iterations to the smallest appropriate size means the quality steps around code being written are smaller and more frequent, meaning they have a larger impact since people are more likely to be thorough and to make changes when they haven't invested as much time. Importantly, Pace teams are not overloading the bottlenecks;

the team is balanced around them, and other capabilities can choose non-bottleneck iterations to fill up their available capacity, or have more readiness for effective handovers.

Now that we've introduced the transition from Sprints to Pace, let's take a look at how some of the Lufthansa Technik rules are applied in Pace for software development.

Similarities between Lufthansa's Rules and Pace:

There are two primary similarities between MRO and software development.

1. The work is consistent in the tasks themselves that are executed. That is, you can define the steps of tasks for a standard operations procedure in an MRO project, and also a software iteration. Also, the variability is high and unknown at the start of the group of work.
2. There is a large presence of highly skilled people: engineers in MRO and software engineers in software. As we know from what's been previously discussed, any environment with a team of highly skilled people will have natural bottlenecking by the skilled people. They are the most skilled workers, so they are the ones most likely to be loaded with tasks; if anything goes wrong or is harder than usual, they are the ones called upon to fix it, and the less experienced workers will load up the experienced team and thus slow everyone down.

So, like Lufthansa Technik, we need to move away from a system that focuses on individual input, and instead focus on team output. To do that, we will need to implement the similar rules, with the mechanics of the way rules function being different for the software environment.

We need to focus on the flow of work through the bottleneck, which is caused by our most experienced resources. Much like Lufthansa Technik, there are huge benefits to reducing work package lengths and the batching effects that go with it, moving our focus to Variation Management of very small, but fast, flowing pieces of work. Let's look at how the mechanics of that can change our current standard practice approach of sprints.

Rule 1: Small chunks of work

Taking away the due dates eliminates the time box aspect of sprints. This allows our now non-sprint iterations to be whatever length we like. More importantly, it allows for Goldilocks iterations: not too long, not too short, just right! Planning iterations to their true estimated time will inevitably shorten the length of an individual iteration as we are no longer fitting it into the artificial time boxes and don't feel compelled to 'add more.' Teams will split what was once a two-week chunk of work into multiple stages because they are the stages that actually make sense. This is done in sprints too, but they tend to get stacked and treated as one iteration. Once this is implemented, teams will naturally start to reduce the time of a planned iteration as they receive the benefits of shorter and shorter iterations: smaller and more frequent handovers, and reviews of the individual chunks. Small chunks of work flowing smoothly and quickly through a team really does feel 'Agile.'

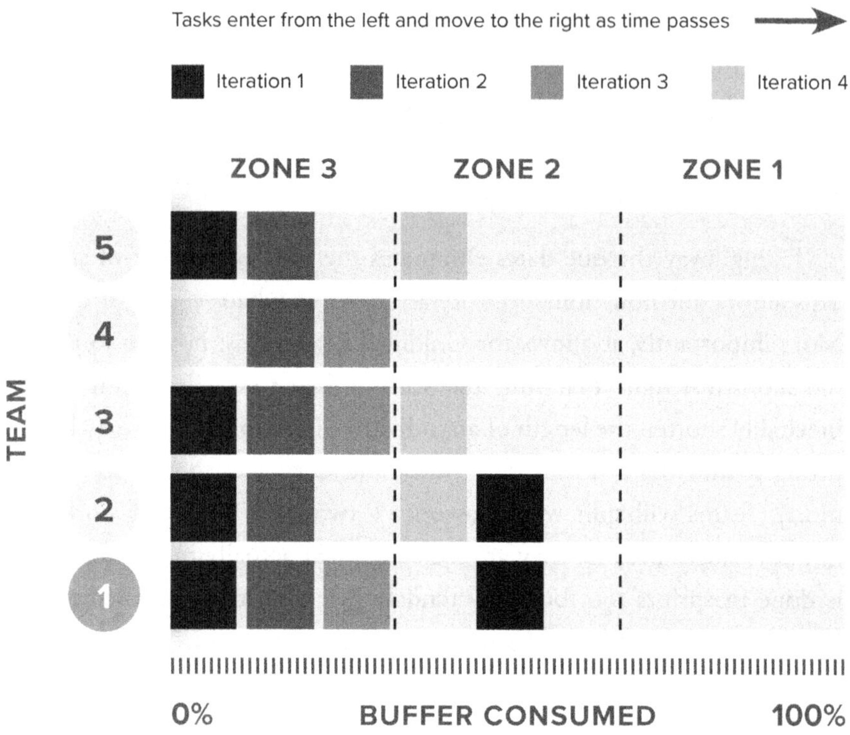

Fig 12 In this diagram we have the same team of five, however, the view and management of their work is now subject to the Flow Management Rules. Their work is now in smaller chunks, with multiple smaller iterations available for the team.

Rule 2: Use time buffers not deadlines

Again, due dates aren't a real project management technique. Some project managers reading this might feel either confused or confronted. To explain, consider this example: having a deadline, or a last straw, and then constantly reminding you of the threat of that last straw is like the tax department threatening huge fines and jail time if you don't file your tax return. Do businesses only worry about their taxes at the end of the year?

Well, some might; but most calculate and take appropriate actions throughout the year instead of waiting until the deadline, or last straw. The tax department threatens with the impending deadline because it has no other way to cause compliance . . . like a project manager who has no other way to motivate a team. This phenomenon falls under behavioural psychology, which shows that small, frequent rewards and feelings of success are the best ways to impel people to do more and more of the right thing. The stick approach also works—and, again, works best when the steps are small and frequent.

You might still be thinking, *As long as they get compliance like the tax department, who cares?* Surging is why you should care. There's a reason surging is mentioned so often in this book: the peak of speed (at the expense of quality) is followed by the trough of neither following the surge.

A system with visibility of progress and an increased risk scale (Variation Management) provides everything we need for a team to perform to the pre-zone 1 target. But we will need to alter the system a bit to really receive the full benefits of progress, along with the frequent small negative reminders of delay, without the surge behaviour at the end. Rather than the visibility of creeping towards the due date, we want the pressure to remain mainly consistent across the days; the goal is to balance out days of surging and days of laziness and instead have a smooth and consistent pace of work getting done. In order to achieve this, the buffer should be extended to the entire window of time. Continuing with our two-week sprint example, we take the buffer sections we built into the last third of the sprint and extend them across the whole timeframe, so we have the 10 days split into 3 Zones. Assuming we're manually managing this, the days in each Zone are relevant to ensure consistent movement of tasks. If it was digital and the tasks advance across the timeframe automatically, we can have the Zones divided into true thirds.

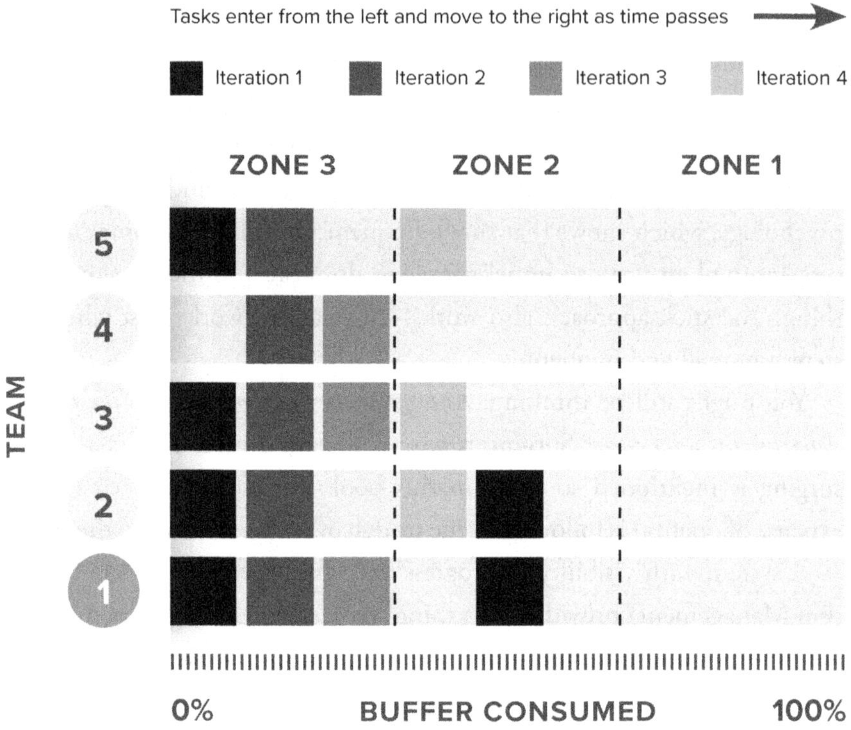

Fig 12 The Diagram shows buffer as a % rather than a timeframe. The Zone states are clear from the position of the task. They do not need to be calculated like we did applying Variation Rules to Sprints earlier.

Now, rather than the progress of time being a line that advances across the timeframe, you must imagine the tasks themselves moving along the timeframe. There should be one day for every day the tasks aren't completed, until they are, and are then taken off the board. This is the same way a Kanban ticket moves from 'To Do' to 'Doing;' these tasks move from 'Day 1' to 'Day 2.'

This seemingly small change opens up huge changes to the management of task flow. A few weeks after making this change, team leaders and business owners are somewhat in awe of the effects (along with the other three Pace rules that come with it). The first benefit is allowing for the elimination of due dates. The majority of work will be started in Zone 3 and completed in Zone 2, which means the average 'completed' line of tasks on the diagram above shows how 'in control' the team is of their workflow. As it moves left, they are getting ahead; as it shifts right, they are getting behind. A line like this acts as the speedometer in the car, rather than an 'over the speed limit' warning light or siren that would cause dangerously drastic responses and overcorrection.

A question this frequently raises in those experienced in Scrum is, "What about all the Scrum elements around the sprint; when do these happen?" To which the answer is, "When do you want them to happen?" To be a little less flippant, these elements like detailing the iterations or running retrospectives can happen whenever you'd like. If you like the frequency you currently have, book a meeting to have the retrospective every two weeks, for example. If you would prefer more frequency, do so. The confines of the sprint deadlines are gone, which means you can change the timings of these things. Of course, it also means you don't have to change them if they were relevant timings. The iteration planning or detailing of iterations is an interesting component—you could do these in planned spots, or you could put 'future iteration planning' tasks on the board with estimates. A third option is to handle them very differently where they are also buffer managed, but in a different way from the way iterations are buffer managed. We will discuss this later in the implementation section, where we design the schematic. The schematic manages all the flow of work from idea inception to completion, with a very late stage being the Pace board.

Rule 3: Limit work in progress

Tasks in play are already a positive outcome of sprints, but the free-flowing Pace system can take that to another level. The fewer things we are committed to, the better; as we are less impacted by external variation, and less work in progress can be active in the team. We want the team to finish what they have started first, and only start what they can finish. The first thing we need to do is ensure there is an appropriate amount of work in play. This is a simple calculation of the capacity of each team member (80 hours in the example containing 10 business days) compared to the available work on the board. As an iteration is completed, that frees up some time for new iterations to flow onto the board. As iterations get shorter and shorter, you will quickly start to wonder why the timeframe is so long, so you can reel in that timeframe to a more useful length. Practically, this usually means taking a Zone's worth of time off (three days in the example), and then of course redistributing the three Zones across the remaining days. Let this settle for a week or two and reassess. Doing this limits the possibility for the team to be working on more things than they need to, which in turn decreases the risk of external variation impacts and decreases work in progress—and the costs of the mental switching that comes with it.

Controlling the iterations in play for the team—being that they are on the board available to be worked on—by the available capacity of all people or capabilities in the iteration automatically ensures the bottleneck is not overloaded. Where you would plan the work for the bottleneck first in a sprint (as detailed in the Bottleneck Rules section earlier), you now allow iterations on the board based on their capacity first.

Choosing the iterations or work to be done is more manageable when you are running this process of deciding what goes onto the Pace board. This is an easy point to glaze over, as most will think they are already deciding carefully what to work on, but the nature of a time-boxed sprint tricks our efficiency-focused mind into making the wrong decisions. In New Zealand we have a lot of small Chinese food restaurants that provide a buffet, where you can purchase a plastic container and fill it with whatever you like. Locally, these restaurants are referred to as 'Chinese Takeaways'. I'm sure it's not a novel concept, but the specific case is etched in every Kiwi's mind. Once you buy the container (pricing is based purely on size) it becomes a challenge—some may say an art—to fit as much as possible into that container. Much like our analogy of the professor's jar full of rocks, gravel, and sand, you ensure the high value stuff is in early (spring rolls and lemon chicken). But the really smart ones figured out you need some rice first, as a base, to maximise the amount you can fit. By the end of this exercise you will stuff anything that will fit into the last gaps of the container (Another cream cheese wonton! More of those little tiny corn cobs!), whether you really want it or not.

A timebox causes the same behaviour—people really want to plan for all the available time, and will throw work in there to get done, whether it's really valuable or not. The free-flowing Pace board takes that pressure off. You could still throw something on that doesn't need to be done, but there isn't the feeling of efficiency loss you get while planning an unfilled timebox, because even if there is available capacity on the board, people still have relevant tasks they are working on. People don't need to be 'full of tasks' to be productive—the board shows that very quickly and clearly.

The unnecessary waste of code that gets half done—or even fully done, and then never used—can be awful. Finance is probably monitoring costs and a few hundred dollars a month wasted will likely get their attention; however, hundreds of dollars of wasted code (that number skyrockets if the bottleneck touched it) never gets seen. Deciding what is worth working on is a huge control point to have. Only commit to valuable work, then get it done quickly with a rapidly moving team.

Rule 4: Have a relevant mix of tasks

You're now making task decisions with increased control and scrutiny, but sometimes you will have the dilemma of deciding which *type* of task should be prioritized. For example, what's more imminently important: bug fixes or paid customizations? This is not the kind of decision you want to have to make on a day-by-day, ad hoc basis; you want this to be a strategic decision made outside of the daily pressures and then executed on a daily basis. There will be an overall mix that makes sense at a team level, or for many businesses at a company level. By having a relevant mix of tasks, the team can give their best to both long-term and short-term goals. The mix of task types might be a combination of bug fixes and roadmap features, or customer paid work. The details and ideals will be different for each team and business, but for purposes of an example we'll say that the mix of tasks will include 25% bug fixes to ensure the backlog reduces. You can provide visibility for that 25% of bug work and write those onto red sticky notes on the planning board. You could allot 50% of the work to the roadmap work, using white notes, and using blue sticky notes to denote the last 25% for building automated testing tools for internal use. Make sure to plan roughly those percentages onto the board, and allocate more work coming in as those iterations are completed. If 8 hours' worth of automated testing is completed, put roughly 8 hours back onto the board. The implementation section has more detail on how you can manage this practically; it does take a little more planning and consideration, but not as much as you might fear.

The final type of task is 'urgent interrupt.' For some teams, managing this properly will double productive output. Urgent interrupt occurs when a business owner or manager has something urgent that needs to be done. We talked earlier about how much damage this behaviour can cause in a sprint. The huge benefit of how urgent interrupt tasks are managed in Pace is that it places the urgent task next on the list, but doesn't 'shove' it onto the board in a disruptive manner. With the rate at which iterations are completed in Pace, that urgent task can be on the board within the next day or so. The team (and the owner or manager) doesn't have to make the choice of waiting until the next sprint, or ruining the current one. This solution is made even more productive by using small chunks of work and a fast-flowing board. This allows the best of both worlds: the 'interrupter' gets what they want, and the team gets to finish what they have started before moving onto the next thing.

Case Study: RockWare

Name changed to protect the innocent.

Objective

RockWare was growing very quickly. At the point at which we started working with them, they were hiring a new developer every week. RockWare looked at where their trajectory was headed, and wanted to implement a solution to avoid hiring one person for every person's worth of productivity they had.

The interesting challenge was that RockWare wanted higher productivity, but not at a cost of quality. Churning out more software full of bugs was not the goal here!

Challenges (Current Reality)

RockWare, being a software product company, experienced the industry-typical, highly-variable environment. Managers reported about 5-6 hours' worth of actual productive time per day from their developers. That was a fairly big opportunity for productivity gain!

Most of the remaining time was burned through the management of work—handovers, planning, teamwork, etc.

The biggest observable loss was the amount of waiting time between resources, i.e. when one team member finished a piece of work and handed it off to the next resource, they'd have to wait for that next resource to be available. This was exacerbated by the fact that the experts in the team were in highly valuable job roles, and presented as a bottleneck for the team; with handovers to or from these experts being critical time-loss opportunities. As a result, RockWare's work in progress was higher than they would have liked.

People felt the pressure to be productive, so they'd go pick up a new piece of work to start working on, while waiting on the bottleneck. In turn this increased multitasking and decreased the flow of the team's work.

Another challenge RockWare experienced was in relation to their review work. In general, the quality of the review decreases as the time between when the work is coded and when it gets reviewed increases, and the situation at RockWare was no different. The issue was that the time between coding and having the code reviewed was too large, and that time period needed to be shortened because it was critical to quality—the shorter the time, the higher the quality!

Process (Solution)

Traditional project management techniques had long been avoided at RockWare due to project management planning being out of touch with software development reality.

In software, estimating into the future becomes less and less reliable to the point of guessing. And committing to those guesses leads many software teams to be compelled by customer due dates and project manager techniques that enforce the wrong kinds of behaviours (e.g. rushing, ignoring quality steps, and cutting back on long term scalable structures).

Essentially, this equates to prioritising speed over quality, which was a strategic deal breaker for RockWare.

The solution implemented at RockWare involved:

- Design and implementation of a visible workflow board (i.e. a schematic of major development milestones)
- Managing work based on true time estimates, rather than an arbitrary week-based Sprint estimates
- Work managers to maximise the value added by expert developers (the team constraints)
- Time buffers to manage risk, rather than internal due dates
- Standups focused on risk management, rather than trivial 'who dunnit' updates

Outcomes/Results

Managers reported that work flowed more smoothly, stress levels went down, and the team leaders and managers could focus on planning work more effectively, as well as getting more of their own development work done.

Upcoming work was visible, so as team members completed tasks, they were able to work on the next item in sequence—this smooth work system created a sense of calm among teams and managers alike.

The amount of unnecessary intra-company emails decreased, because the necessary notes were on the tasks, the relevant tasks were on the board, and the documentation made for clear trust and communication amongst staff and managers alike. Managers no longer needed to chase progress of tasks or projects, as all was visible on the newly implemented dashboards.

Using this system helped RockWare clearly know their own pace, what the team could accomplish, and how long it would take. This was fantastic for team members, project managers, and the C-levels alike.

Management noticed that team members quit panicking over lack of resources because they now had clear visibility of the organisation of the workflow, and clear visibility of the available capacity—that used to be hidden by high WIP.

RockWare went from hiring roughly a person a week down to a few developers over the entire year. The project paid for itself in manpower immediately. Once the initial team saw success with this process, it was rolled out to the rest of the company.

RockWare is committed to ongoing process improvements and continuously tweaks their standard operating procedures, and these changes are immediately implemented across the whole company, as all teams run the same system.

CHAPTER 11

A Good Day in the Life:

Dave can just feel it—he's about to have another good day . . .

12:15pm—He's been at work less than an hour, and he's already dealing with an emergency. Sarah, one of the product owners, is on annual leave this week, but she neglected to tell everyone (or just forgot) that a bug fix was promised to a major client by Monday. Unfortunately, it's now Tuesday morning; however, the bug was fixed last week as it flowed through the schematic after she created the iteration and put it onto the Pace board. Fifteen minutes ago, Dave took a frantic call from Derek in customer service, who was demanding to know what was going on.

"I have the customer waiting on the phone right now, and they're NOT happy!"

Dave begrudgingly promises to look into it immediately.

"I have absolutely no idea what's going on—it's Sarah's job to deal with customers, not mine," Dave tells Ritika, a senior developer. "And the team's capacity is accounted for in Pace. I'm not breaking the rule; it's working so well."

"Calm down, Dave, it was done last week; the task was on Sarah to update Derek, which we reassigned today at the Standup where we caught it. I'll shoot it over to Derek now," Ritika says.

Dave sighs, "Okay, sorry, old habits die hard . . . thanks."

1:30pm—With the earlier emergency long dealt with, Dave has managed to catch up on his Slack messages and do a few designs for the team. He's just settled down in a bean bag for a coffee break to skim a few Stack Overflow threads, when he's interrupted by an email alert from Derek.

"An urgent feature request has just come through from sales. Can you get the design spec'd today so I can get the signoff from the customer? Jenny over in sales promised we could get the design sorted today. If we drop the ball on this one, it won't look good."

Dave puts down his cup of coffee and sighs; he wishes sales would stop making promises like that. He responds to Derek's email: "Put it on the Pave board as a Gold task, this will ensure it flows onto the board next, we're doing pretty well so we should be onto it after we finish the current iteration this afternoon. Take a picture, show Jenny, and remind her not to make promises that are shorter than our buffer length."

Five minutes later, Slack chimes with a notification from Jenny: "Thanks for getting to it so quickly. I actually promised the customer it would be done this week, but wanted to give you the hurry-up. Sorry, old habits…"

5:25pm—Dave has his fingers crossed; the last half hour has been free of interruptions. He's been able to complete a chunk of an iteration, and even review Sam's code and take the time to talk through the review with him. If he can keep this up, he'll finish work on time again today. He's about to move to the next thing on his board when the phone rings; it's the accounts department.

"We have a minor invoicing crisis," Henry from accounts says. "I think we figured it out for ourselves, just wanted to check our assumptions."

Dave chuckles. He can tell from Henry's tone that Henry has solved it and wants to show off rather than actually check. This would only take a few minutes, and Dave figures Henry would be more chuffed in person. "Assumptions? I'll be right there."

7:45pm—Dave's been getting things done all day. Not that the team has 'urgent' tasks anymore, but they've even found time to mostly complete the surprise Gold task from earlier. Ritika drops by his desk on her way out the door.

"Have you been able to review Sam's code yet?" she asks.

"Yeah, there are a few things he needs to work on, but he's coming along really well with the extra time I've got to actually coach him now," Dave answers without turning around. "I've been on fire today, solving people's problems left, right, and centre."

Ritika raises her eyebrows sarcastically. "Oh, you've been the hero? From what I can see, everyone is taking care of themselves just fine. I hope you're not one of those hair-on-fire managers that craves a crisis . . . let me talk to Chris about it. Maybe there's some drama we can create for you." She smirks.

Chris

If you think Dave's life looks good now, make sure you check in with his boss, Chris. Chris is the hardworking owner of the company Xyphyr. He works long hours because he loves it. Almost any other software company would be struggling to keep up with all the rising competition. But Chris is getting further and further ahead; with the teams all but on autopilot, he's got the time to work on the business and strategy. Chris thinks, *We have options—we either increase prices by adding more valuable functionality, or we expand our offering to include another niche in our target market.*

In addition to dealing with unexpected emergencies, Xyphyr's teams also work directly with customers to understand their business and create high value functionality. Not only do those new customers want bug-free software (which they more or less get now, certainly by the industry's standards) but they also demand ever-increasing amounts of new features and customization. One customer demands tailored calendar integration for its HR processes. Another wants specialized email tracking. And yet another demands accounting integration from an obscure company.

This not only adds to the value of properly understanding why they want those things and building solution-based features, it also increases the skill and engagement of Chris's already overperforming development team.

And the frustrating thing is, no matter how much "hoop jumping" the customers think Chris is doing by 'expediting' these features that benefit multiple customers, his clients show little gratitude. They'll probably be back on the phone in a week, demanding another thing or complaining about how the price is three times that of the competition. *If only they would remember the software price pales in comparison to the value it brings; oh well, I guess they have to do their due diligence by at least asking for a discount.*

It's this kind of thinking that is currently the subject of a friendly discussion between Chris and Dave.

"We need to target this new niche," Chris suggests to Dave. "Definitely."

"Chris, I'm telling you—we don't have enough understanding of that industry just yet," Dave insists. But we're caught up on the bug backlog and well on track on the roadmap, so we can probably have Sarah get stuck into it as a serious project when she's back. Dave can run the team, it's certainly not a two-person job anymore."

"Well, that will change once Jenny starts closing clients because we're dominating what InstaHR or FaceHR are now offering to every client," Chris points out.

"Hire again?" Dave suggests.

"If you'd seen the P&L for the last six months," Chris says, waving a hand at his monitor, "there's plenty of room for it, we've got serious cash to invest. We have two tools in our kit—a gold mining pickaxe and a gold pan. That's all we need; we don't have to worry. We drop prices as we increase the value of our offering and we capture more and more of the market with our expanded offering to more industries. And we can't do any better on expanded offerings than we're currently doing; we've had the productivity gains from Pace, and we've scaled up a lot considering we've not had to hire a developer in the last six months. But if we don't start hiring and perhaps even create a new team, we won't be able to tackle the new areas of business as quickly."

Dave is lost for words; not long ago, he was wondering how long the team or the company would last. He knew that it was wrong to add a whole different industry to her already overperforming development team—they're doing great and there's no reason to ruin that. Intuitively, he feels that it would be a risk adding a whole other team, but that might be old intuition kicking in. As he had told Ritika, old habits and old intuition die hard. He also knows that Chris understands this, since he's had the time and techniques to properly consider things.

Meanwhile, Chris disconnects his laptop from his monitors and slides it into his leather satchel. Swinging it over his shoulder and heading for the car, he can't help but once again think about how different things are now. He and his co-founder often discussed their vision and what it might feel like. Six months ago they were merely talking about it, a few months ago it feels like a blurry dream, and now they are living their vision: having teams that are capable of looking after themselves while Chris can focus his time on the strategy of the product. He's able to meet with customers, planning the future features pro-actively rather than reacting to demands, and handing over a high-level roadmap to people like Dave to get on with delivering the features however he finds best—actually utilizing his strengths in dev design, accelerated by new thinking and effective techniques. What a wild ride.

PART 3

PACE IMPLEMENTATION GUIDE

CHAPTER 12

Solution Design

The first step of solution design is to define the schematic of workflow through various teams and individuals. This already exists in your business; however it may not be documented anywhere. At the risk of sounding like a Chinese proverb, the natural flow will be there, like a river carving its path down a mountain. If you look at a river though, you can see the water takes the easiest path at every turn—there's no long-term planning, there's no thought given to where the endpoint is or how to get there, because water can't think. Ideally, this process should be less like a wandering river and more like a Roman aqueduct. You'll have to put in the effort to haul a bit of stone around and cut through some mountains, but the end result will be the water (code) getting to where it needs to be—in not only the best method, but also a smooth and steady way.

To design a schematic, you need a few components. You need something to show flow, for which we will use arrows. The flow goes from top left to bottom right, meaning arrows point vertically down or horizontally to the right in most cases for normal work. The only situation it flows the other way is accounting for rework of some sort. Our vertical axis shows the progress between the components, and the horizontal axis shows elapsed time to work on a task once it is given to the team.

The work moves through these components as iterations. Just like a sprint iteration, these iterations can contain multiple tasks which should all be completed in succession (likely by different people) once the first task is started. Also like a sprint iteration, an iteration may be a complete piece of work to be released into the product (like a bug patch), or it may be one of many, as part of a larger piece of work (like a feature). We will go into a lot more detail about iterations in the Standard Operating Procedures section; for now, we'll simply understand iterations as they relate to managing schematics.

The two main components of a schematic are 'buckets' and 'buffers,' and the way work moves between them is dictated by 'transfer rules.' Buckets hold work, in various states, that is not yet available for a team to work on; these keep similar tasks together to be managed. Iterations progress from one bucket to the next once they satisfy the transfer rules. For example, a rule might state that no task can take 8 hours or longer, meaning that any longer tasks will have to be broken down into smaller chunks. If a 12-hour task is broken into 3 x 4-hour sections, the iteration containing that task can then move to the next bucket since the rule is now satisfied. Transfer rules also allow iterations to flow onto the buffer, or direct them back to an upstream component. Remember from our Pace section, the buffer is where iterations are worked on by the team, in much the same way a Scrum-based team would work on sprints. The target for the team is to not let iterations use the whole buffer—meaning the iterations take longer to complete and move out of the buffer than the allotment of the buffer's length. This is essentially the game *Space Invaders* for task managers, with a few developers sitting at the bottom firing code (or any work) up at the tasks as they keep moving closer.

The following is an example of what a sample schematic would look like for a simple software team:

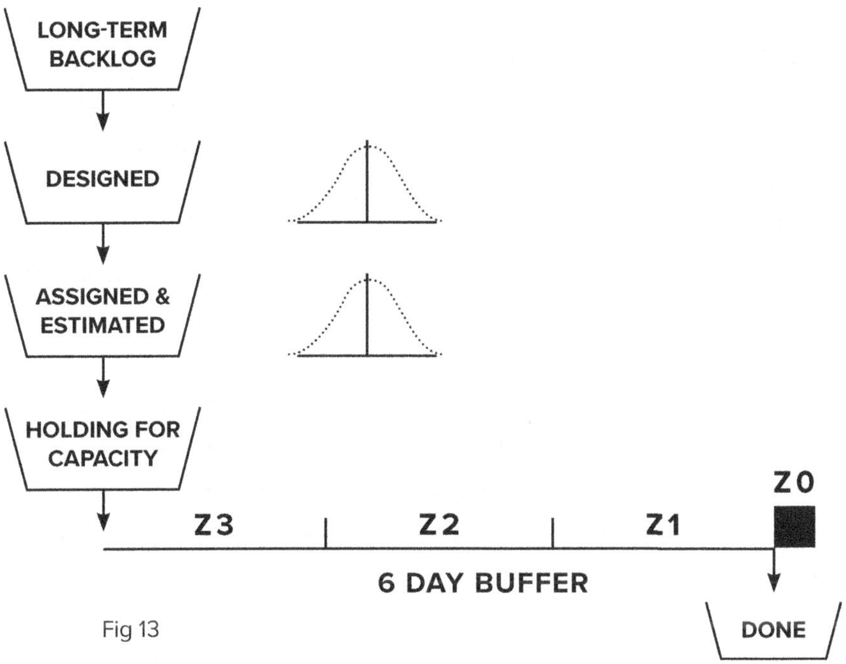

Fig 13

This schematic includes four buckets, each with transfer rules between them and the next component. Two of the buckets have acceptability bands, which are a visibility system to see and manage how 'full' they are—there should be enough work in there, but not too much. Then there is a 6-day buffer, which means the team has 6 days to complete all of the tasks in any iteration which is in play for the team. Once they're complete, they are moved to the 'done' bucket, which holds finished work. If there were more teams using Pace in their business, moving the iteration to 'done' would trigger the handover of the iteration to the next team. If they aren't using Pace, it will be handed off in the currently used way, whether that's an email or other method.

Once iterations are in play on the buffer, they are managed by the team themselves via the 'Buffer Board' which we cover in the Visibility & Execution section. For now, we'll focus on the buckets.

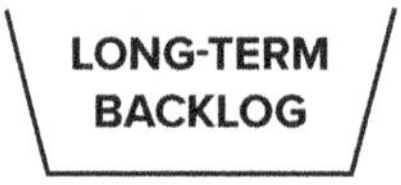

Fig 14

The first bucket for Xyphyr is 'long-term backlog,' which should be familiar to any software team. This serves to hold anything that has been created as an idea, which can come from many different sources. This might be bugs which have little impact but have been identified, feature requests from customers, roadmap items that need a holding place, or various dev ideas. It is the catch-all funnel at the start of the schematic, which captures any and every piece of work for the team.

Fig 15

In order to leave the bucket for the next 'design,' the iteration must satisfy the transfer rules. In this case, the rule states that an iteration has had a high-level design (HLD) done. In Xyphyr, the CTO does the HLD, so it is just a bucket. If it were done by the team, it would be released to the buffer instead.

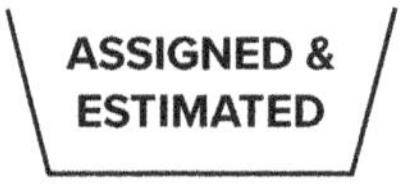

Fig 16

The next bucket, 'assign and estimate,' is fairly self-explanatory. The rules require that the tasks in the iteration each have a resource assigned to them. In a few cases this will be a person, but in Xyphyr's case the resources are capabilities, as we discussed earlier. Those tasks also need to have estimates assigned to them. If they are template-based, most of the assigning and estimating will have been pre-populated (see SOP section for more).

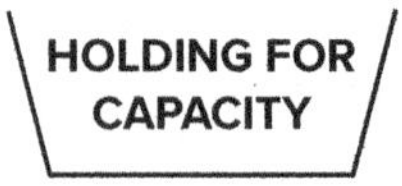

Fig 17

The final pre-requisite bucket is the 'holding point for capacity'. This is where all the previous pre-requisite transfer rules are met, but there is currently no capacity on the buffer to pass the iterations to the team to work on.

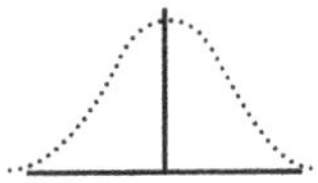

Fig 18

The Design and Capacity buckets have acceptability bands. Acceptability bands have a low danger level through to a high danger level, with the goal to keep them in the middle. In the same way, a temperature gauge on your car's dashboard functions as an acceptability band. There is a range that is acceptable, and risk rises as the temperature deviates further from the midpoint. If we take the analogy to the thermostat for the air conditioning in the car, the people responsible for the acceptability bands should react like the air conditioner: steady when close to target and drastic when things are well outside of acceptable range, like when your air conditioner blasts to quickly cool a car that's been baking in the sun on a hot day.

We have these buckets as a form of queue control to ensure there is enough work ready and waiting for the team to work on. We also don't want these buckets to get too full because priorities are constantly changing in software development, along with work approach and many other factors. So there should be enough work spec'd to develop, but not too much high-level designed that could become stale by the time it gets to the team—this is the far queue of work.

The near queue of work is the capacity bucket; we don't want this empty for obvious reasons, but we don't want it too full for management reasons. This bucket is where the team leader or nominated distributor of work sequences the work to be done. If there is too much in this bucket, the leader or distributor can waste a lot of time making incorrect or inconsequential decisions. It's important to work on sequencing the tasks that are 2nd or 3rd in line, but not those that are 56th or 57th. The sequencing of work has its own set of rules, to both make sequencing more intuitive for the sequencer, and to ensure mix rules are followed. These rules should follow the best practice decided for the team's effective mix of completed work. Our example has two rules. The first rule is to maintain about 20%

work as bug fixes. To follow this, the scheduler just needs to look at the work on the board and release either bug or non-bug work based on the current work on the board. A simple way to maintain this is to schedule in what came out, which needs to be checked against the actual work on the board from time to time as not all iterations are the same length. Our other rule is 'executive override' for urgent jobs, a rule that is often required as a team is first introduced to Pace, but becomes used less and less as execs begin to trust the system's reliable speed. The executive override[4] allows someone who, under the previous method, had the authority to interrupt staff to put something at the front of the queue. This allows for urgent attention while not actually interrupting the team; the overridden task will just be first in line once capacity allows. There can only be one executive override at once, so the execs must choose which of their many important things really matters most.

The total sum of transfer rules should ensure that only work that can be finished is transferred to the board, and the sequencing done at the near queue should ensure that work which is started on the board isn't moved or re-shuffled. Transfer and scheduling rules can be as complex as the people running them can handle—simple rules are recommended, as the benefits of additional rules diminish while the complexity of multiple rules compound. In this case, software can allow for complex rules that apply themselves automatically and consistently.

To recap, let's follow a new task through the schematic. The CTO, after meeting with the team leader and discussing the items in the 'long-term backlog,' decided there's a simple feature—a 'last used function' button—that could be added to make the home screen of their application more user-friendly. That item, along with many others, is moved into the 'design bucket' because it has met the transfer rule of:

---

4. Xyphyr calls this a Gold ticket, as mentioned in Dave's 'Good day in the life' story later in the book.

✓ **Planned for development**

A few days later, the CTO picks up the item and writes the high-level design for the feature (since they are a small company, and she also does the high level designs); once done, she puts the task card into the 'assign and estimate' bucket. This is because the task met the transfer rule of:

✓ **Designed**

The team leader picks up the item from there and fills in the iteration details for it, assigning the tasks to the relevant capabilities. Since this is a straightforward task, he uses some standard estimates (based on brackets); if it were more complex, he could ask the people in the capability to estimate the task at a weekly meeting. The iteration then moves to the 'sequencing bucket' because it has met the transfer rules of:

✓ **All tasks assigned**

✓ **All tasks estimated**

The next morning, the team lead is looking at the upcoming sequence and decides to assign this iteration to spot number 7. A few days later, the iteration has made its way to the top of the list. One of the devs takes a freshly completed iteration off the board, freeing up enough capacity for this iteration. After noting the iteration removed was also feature work rather than bug work, he moves this new iteration onto the board, because the following rules are met:

✓ **Sufficient capacity available for everyone who needs to work on it**

✓ **Correct mix between bug/non-bug work**

Another few days pass and the iteration is successfully completed in Zone 2 of the buffer. The iteration is taken from the board and put in the 'done' bucket.

✓ **All tasks complete**

CHAPTER 13

Creating your schematic:

Scope

First you must define the scope of your team.[5] This will be based on whether your teams are organised by department or by product. If you are department-based, you should first identify the bottleneck department, as this is the team that will provide the most value from an implemented buffer board. After the initial install, you can introduce buffer boards to other teams to ensure their work is flowing smoothly and visibly, thereby reducing the risk of it impacting the constraint. Until then, buckets can be used to hold work as it flows from department to department, and acceptability bands to ensure there is sufficient work for the bottleneck (and also to ensure WIP doesn't climb). If you are running a product team, you also need to decide whether it is just a development team, or whether planning work is also included. If, like our example above, you have a founder or executive completing a step of the process, it's strongly recommended you keep that work in an acceptability-band-monitored bucket. This maintains the level of work in that bucket and gives everyone a simple way of seeing where it is; for example, you might top it up to a high level before the exec goes away on holiday, or use the dangerously low level to poke them into action. In our experience, execs don't like working to a task board and will break the system if they are asked to.

5. Each system has a bottleneck. The system you are implementing in will define how detailed to go. For example, a business could have 'development' as the bottleneck, and within development 'testing' could be the bottleneck, further the testing team could have 'Ben' as the bottleneck.

In most cases, starting with just the dev team is the safest approach—it simplifies the interactions between teams and management in the system. Once you have the dev teams working effectively, surrounding work could then be applied to the buffer boards as well.

Flow

What are the major steps your work goes through, from inception to completion? Start by tracking down all the sources of work for your team.

You may think there is only one place or a couple of places that work comes into the team from, and in some cases you will be right. In many cases, managers are surprised by the number of different systems in play to both provide a source of work or a way to store and manage it. You will likely have one or two primary systems, such as a customer service ticket system, and a dev team ticket management system. Other systems may include people external to the team who add work via email, Slack, a phone call, or even personal handovers. Once the work is there, some will formally add it into the system as they should, others will keep track of it in their head, there might be a small team that uses Trello instead, another individual may use Excel to track their work, and finally a manager may use a personal notes app on their phone to keep track of things. Having all of those systems simultaneously is the worst case scenario (and I've seen scenarios very close to this!) but you will need to identify those you do have in your team. The 'processes' that are actually in place are rarely those that management thinks are there.

In order to identify the current systems, you will need to talk to individuals in the team, ask about the work they are doing, where they got it from, where they're keeping track of it, and where it will go once they are done with it. Finding the path for 50 to 80 percent of the work will be done relatively quickly this way. Digging for the rest will take time; depending on the level of difficulty, you can begin with knowing 95% of the work systems. The remaining 5% will become obvious once you implement the new system.

Map all these pathways in a simple flow chart. There are many different methods of drawing these flowcharts—either use one you're familiar with, or just make one up. As long as your symbols are consistent, it doesn't matter; this is just to create a draft of current practices. A simple example looks like this:

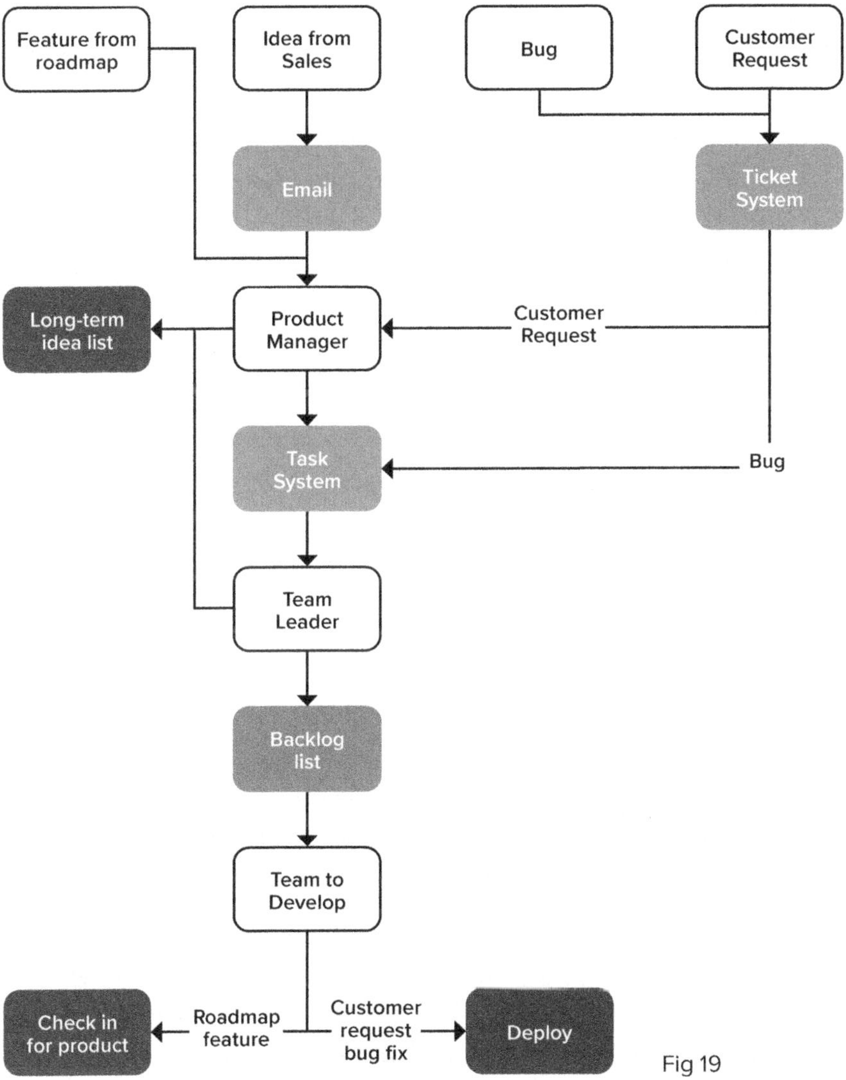

Fig 19

To test your estimated current flow, you can 'tag' a task just after creation—from whatever creation point you are testing (e.g. customer bug)—and ask the person who first touches it to let you know when they hand it over, and to whom. Follow that around the business until the task is closed as complete. That should confirm or correct your draft of how work flows and is stored. It will also give interesting information about how long it takes a task to travel through the system. While people will likely expedite that particular task since they know you are tracking it, there is a more detailed step you can do if your flow is currently well established in systems. Find a new ticket, take a note of it, and discretely follow the process above to see how long it takes to get from place to place. Then follow the process above, but ask everyone to treat the task as their next priority. That is, to finish what they're working on now, then move onto the followed tasks next. The difference in time between the first and second task will give you an indication of the effect of work in progress on the responsiveness of your system.

Once you have a flow chart containing all of the entry points of tasks for the team, and the flow of the tasks, we can move onto applying buckets to the schematic.

Buckets

Buckets provide management points within the system. Work at similar stages should be held in the same place so each stage can be managed more effectively. A Pace system works as a pull system from the team. Earlier we reviewed the benefit of starting work as late as possible to minimize the impact of external variation; the same concept applies to these buckets. We only want to move work through the system as it is valuable to do so. We need to ensure our teams always have enough to do; however, as work flows down the buckets, more and more time has been spent on it, so the potential for wasted time is greater from variation impacting these tasks.

The exception to this is the first bucket, which is called Entry. We want everything to be created there as soon as possible, because the alternative is to have it remain stuck at the previous step (such as the customer's mind, an email, a scribbled note on a desk, or a notebook after a meeting). Entry tasks should be created in the system so they are captured somewhere within the process.

Create the Entry bucket (you can call it whatever you'd like), and draw all the 'inbound flow lines' pointing into it. You now have one united flow. Continue with this primary flow for now, and then come back to any exceptions at a later point. Entry flows into the team's work. You will need to discover the logical steps that you wish to use as control points. As with our Xyphyr example, there are some common starting points to consider. Generally, you will need somewhere to hold long-term work that you've decided to deliver, then a shorter-term backlog where you will spec the work to add more detail, and then you will need to set the size of the iteration. Once the iteration is set, you will want to assign and estimate the individual tasks in the iteration. Finally, work in the final bucket is pulled into the team as capacity allows.

Following this path, decide if you want to separate 'all work' from work you accept, but don't want to plan yet. If that is the case, you will need a separate 'long-term backlog.'

You will likely need a short-term backlog of some sort, to separate out the work you have agreed to queue up for the team. Decide whether you want the work to be detailed before it is allowed into this bucket, or afterwards. If afterwards, create another bucket labelled 'Designed' or 'Spec'd.'

Here is another question of granularity: do you want the work to only be labelled 'Designed' once all tasks are assigned and estimated, or would you prefer that to be another bucket?

The final bucket before the Buffer should only contain work that is 100% ready to work on; this bucket is merely a holding tank for the team, ready to be put into play on the board once capacity is available.

How do you know whether to include all or some of these buckets, or add more? Remember that an increase in control and visibility comes at the cost of more time spent managing the process. Fewer buckets will mostly suit smaller teams, while more buckets will suit larger teams or a greater number of teams. Keep in mind that a dedicated bucket for external project factors may need to be added. In our Xyphyr example, we had a bucket just for designs because that included someone external to the team. Your business structure will determine whether or not you need this—is there anyone outside the team that needs to have input into the work? Some examples of these external factors could be: customer service team, implementation consultants (either within your business or external channel partners), sales people, or internal 'customers.' You'll want a dedicated bucket if similar factors are involved so you can apply the acceptability bands, which will be the next step.

Once the primary flow is in place, go through each of the additional flows you identified in the flow mapping exercise, and consider whether each flows through the primary buckets or deviates at some point. If it deviates, should it? Is it valuable to do so, or is it just the way it has worked out? Decide where you want each flow to travel. Repeat this exercise for each additional flow. If you have the same number of deviation buckets as primary buckets (or more), you should review again. That scenario adds a lot of complexity, and is a very rare circumstance. The probable cause of this is trying to accommodate too many people's special cases that, in all actuality, add nothing to the process. Deviations should only occur where they are clearly necessary; for example, the external parties involved in the process listed above. Don't accommodate the process of the old dog that's been around forever just because they're a grumpy bugger and you want to avoid a painful conversation.

Acceptability Bands

The next step is to decide which of the buckets need acceptability bands. We've already discussed the purpose of acceptability bands; now we'll explain how to decide which buckets to apply them to. This decision is based on how important it is to ensure the bucket doesn't run empty and how high the risk is that the bucket might run empty. For example, it is very important that the final bucket before the board doesn't run out, as this is what feeds the team, so we always use an acceptability band here. On the other hand, the entry bucket doesn't matter as much, since you can easily find work to fill it if you need to; I've yet to see a software team that has no idea of what to do!

The buckets at the highest risk of running dry are the buckets we mentioned earlier where an external party is required to be involved. You will need a clear system (such as Red, Amber, Green) to let you know if you need to increase your activity to get more work ready in that bucket, or if you need to increase the pressure on the external party. We've found that externals love being kept up to date on the state of the acceptability band, too; it works just like the buffer, where it allows them to plan their time interacting with the team appropriately.

You can apply the 4 Rules for Variation Management to these buffers too, the degree of detail you will go to will be based on how available this external party is.

Transfer Rules

Defining the buckets gives us an indication of what the transfer rules will be. The value in having the separate buckets is that the transfer rule is essentially the arrow on the diagram. Because not all schematics will be simple, where all work progresses from one bucket to another, it is important to have the transfer rules that clarify, "If these rules are true, go to this bucket," or, "If this separate set of rules are true, go to a different bucket."

The other reason is that, even if we have defined our buckets and the schematic is a simple single bucket-to-bucket arrow, the transfer rules allow us to define the specific conditions that must be true. You have seen this earlier, in the Xyphyr example where we defined the specific rules.

The rules will depend on the buckets and the arrows you have defined in your schematic, but the way the rules are written are very specific. If you're not a developer, you may wish to have one help you with these, as they end up written like Pseudocode. The rules should be written as 'true' statements—if x is true then transfer—you can and often will have multiple conditions to be satisfied before the iteration can transfer between the buckets. For example, "if all tasks have capabilities assigned to them, and if all tasks have estimates against them," then the iteration can move into the Assigned and Estimated bucket.

These rules can get immensely complex, especially if you introduce 'or' conditions. The natural back pressure on this is the complexity of figuring out what can move in a manual system. Transfer rules are one of the most powerful advantages of using software to manage Pace. As you can imagine, you can make the rules as complex as you like; this won't matter at first, but will make it increasingly effective as you tweak and improve your schematic with time. Of course, the software will not only run these rules exactly as defined every time, it will also automate the application of the rules—meaning the running of the schematic and the following of the rules doesn't require a person to do so. People can just do the tasks as they are required to do, and the iterations will move on their own. I promised to only give solutions in this book that you can apply without software, so I won't go any further into this, but you can contact me for information about the software we use if you are interested.

Buffers

Defining the buffer feels like the most important step in all of this (and it is), which may give you paralysis by analysis (and it shouldn't). Due to the nature of a buffer being there for flexibility, it is easily adjustable. If you make the buffer too long, you will find too many iterations in a state of 'started in play.' This will increase multitasking to a point that is not ideal, bringing you back to pre-buffer status. On the other hand, if tasks are filling the entirety of Zone 1 and still aren't completed (we call this Zone 0), or no one other than the bottleneck has anything to do, then the buffer is too short. In either case, it is easy to adjust the buffer up or down.

If you wish to use your sprints as a reference, it is even easier—you can start by using buffers that are close to your current sprint length. I say 'close,' because, if you are using a manual system, a 9-day buffer is easier to divide and manage over 3 zones than a 10-day buffer, and the one day difference is neither here nor there. You should set the expectation of the team, and yourself, that you will reel in the buffer length shortly. If the team follows the Pace rules effectively, you will quickly find you can reduce the buffer length by a third, and another third not long after.

If you are not running sprints—or are sufficiently convinced that sprints are always an arbitrary length and want nothing to do with them in the new world of having a team working at Pace—you can use the method I described in defining the flow of the schematic. Rather than for the whole schematic, we will just use it for the tasks in play for the team. This is a fun exercise. For the first test, tag an iteration (don't tell the team which you have tagged) and note how long it takes it to travel from the 'to do' list to its finished state. Once you have run that test and have your answer, tag another iteration—but this time, tell the team to treat it as the highest priority. They shouldn't rush it, but should prioritize that task when it is handed to them. You will now have your 'normal lead time' (which includes the impact of WIP) and the 'true lead time' (which is how long it actually takes from start to finish). An interesting third step for the team to do is compare that time to the task estimates, as it should be the same as the sum of the estimates. This shows how accurately their estimates match real-world task work which accounts for Grey Time.

To acquire a surprisingly accurate starting point, halve the length of your normal lead time iteration, and double the length of the true lead time iteration. Use the longer time as a starting buffer, to be safe.

CHAPTER 14

Mapping iterations and standard operating procedures

Now that you have built a schematic for your iterations to flow through, you will want to ensure that the tasks completed within each iteration are the best tasks for that iteration. This is a standard operating procedure and ensures a higher degree of uniformity. This isn't intended to stop the team from being flexible in how they do tasks, especially different types of tasks. If you ask 5 different team members what tasks need to be done for an iteration, you will get 6 or 7 different answers. Some of those will be examples of different circumstances, but it also stands to reason that, for many iterations, there is a best base set of tasks, and something (usually quality) suffers if they aren't completed.

This forms a standard checklist of tasks to be completed for each different type of iteration. It is easy to fall into the trap of thinking people know what they're doing, or assuming they will do every step every time—but when they take longer than expected on a task and really blow out the estimates, will the team still take the same time on all the quality steps? Some teams will, but some won't; some will in some conditions, and not others. This step is about creating the templates of the tasks that will go into each type of iteration. They can be adjusted to specific cases, but this ensures there is always a solid starting point.

The degree of detail you go into will depend on how different the iterations are. There will be a different set of necessary tasks for a bug fix than there will be for a new feature from the roadmap, which will be different again from a customer requested customisation. If you choose to dig into this further, you can have multiple templates for any of these types of iterations. You could have templates for different types of bugs, bugs from different customers, different service level agreements, etc. Again, the back pressure is managing all of those templates, and this is an advantage of a software system to support this. You will find in most cases that the different iteration templates that are best for a team are the same as the different types of iterations chosen at the work mix management stage. If the team has maintenance iterations, bug iterations, and server optimisation iterations in their work mix, those will be good starting points for the templates.

Assuming a manual system once again, your iteration card's task list will look something like this:

Done	Task	Capability	Estimate
	Design reviewed	Dev Lead	0.5
	Research	Developer	2
	Coding	Developer	6
	Testing	Developer	2
	Testing and Review	Senior Dev	1
	Review	Dev Lead	0.5

Fig 20

To create each template, follow a similar process as mapping the workflows in the Schematic section. In fact, you may have gathered sufficient information in that step to put together some draft templates. If not, ask the team what steps they feel should be included in the given iteration type. You may need to badger people somewhat to ensure they are accurate and complete. Once you have the information, you can have the steps printed off and ready to go, with spaces for any additional information you want to include (e.g. the ticket number from your ticket software or ERP system).

Ticket #

Customer or Project

Done	Task	Capability	Estimate
	Design reviewed	Dev Lead	0.5
	Research	Developer	2
	Coding	Developer	6
	Testing	Developer	2
	Testing and Review	Senior Dev	1
	Review	Dev Lead	0.5

Fig 21

The final card will look something like the one above. Keep the card as simple as you can while achieving the desired effects. I used handwritten file cards as a way to get started on more than one project; it's a lot easier to adjust when you can just scribble something else on the card if you suspect the template will change once the project actually begins.

A final note: It is unlikely you'll want to create a card for every single thing that lands in the entry bucket, so you will need to decide the point at which you switch from a simple list or sticky note system to iteration cards in a bucket. I would usually suggest creating the card as the detailed design for the iteration.

Executing day to day with a visual dashboard

We have the schematic to manage the flow of iterations, and now we have the task cards to actually flow through that schematic . . . Now we need a dashboard (or two) to give us visibility of the system. You *can choose* to show buckets on a dashboard, and you *should* show acceptability bands on a dashboard, but you *must* show the buffer on a dashboard, so I'll start there.

Buffer Board

Our buffer board will look very similar to the diagrams in the Flow Management Rules section. The board should be divided into rows and columns to represent Zones on one axis and Resources (being people or capabilities) on the other. It doesn't matter which axis is which; likewise, it doesn't matter which direction the iteration cards flow (from left to right, right to left, bottom up, or top down). We tend to use the X axis for resources (along the top of the board) and have the cards flow from the bottom to the top. Considering that the buffer boards often use landscape orientation, this method allows the best use of space, accounting for multiple resources across the top, and utilizing landscape-based cards.

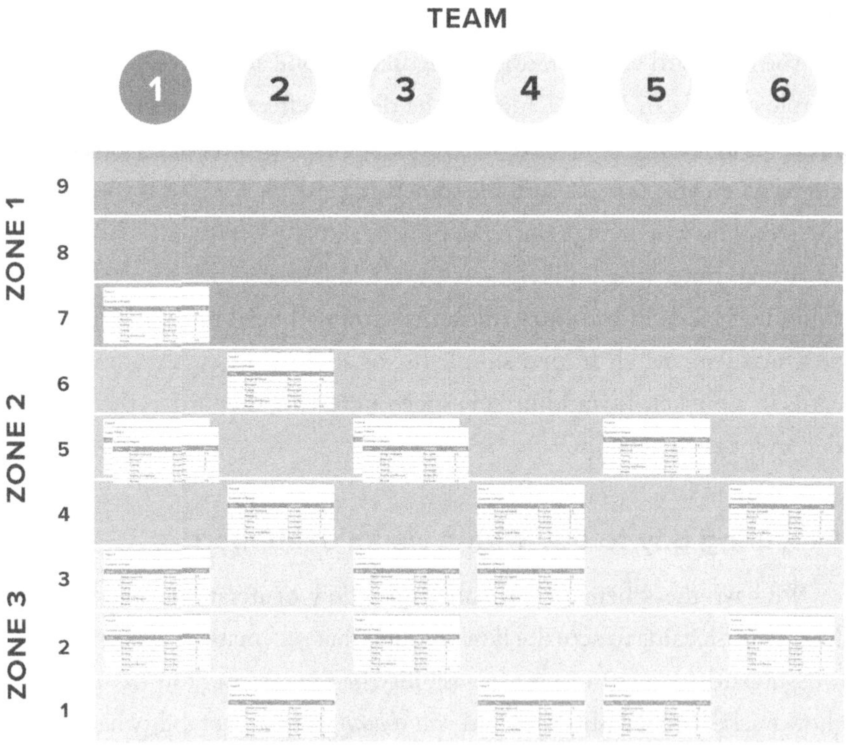

Fig 22

Of course, if you prefer your board to run top to bottom instead, you can get a neat *Space Invaders* vibe going, striving to eliminate iterations before they get to the bottom! The most common case people make for a horizontally flowing board is when they have a large section of wall dedicated to the board. A whole wall board is great if you have the option, as it really draws people into it. The more the team makes the board their own, the more accurately they'll use it. People take care of something they have pride in—I've seen some really customised boarding, the best being a very detailed Mario Kart-themed board. In the Implementation Tips section, you can read about a few more of these tips regarding team involvement and ownership for an enjoyable solution.[6]

The buffer Zones should be separated into subsections—we target 3 subsections within each zone. These subsections show how far a card moves at each update. The purpose of these subdivisions is to ensure a consistent progression of the cards, and it also gives greater visibility of the passage of time, enhancing the behavioural effects of this discussed earlier. Transitioning a card suddenly from Zone 2 to Zone 1 doesn't allow the opportunity to get these benefits. We usually aim for 3 subsections because of practical reasons—a 9-day board divides easily into 3 subsections. If you have a 5-day buffer, I would suggest making 10 subsections: 3 for Zone 3, 4 for Zone 2, and 3 for Zone 1.

As you may have noticed, you should increase the frequency of the updates as the timeframe of the buffer gets shorter, as tasks and iterations move faster we need to ensure we still have time to react to at risk tasks. The more frequent the update, the earlier we can react. Within the 5-day buffer, the team would get most benefit from running an update twice a day. This will sound excessive if you've never been part of a sharply run daily Scrum or other type of stand-up meeting. See an example of how this is done in the Updates section below.

6. Consider this a challenge for anyone who implements Pace. Send me your best and see if you can top the Mario Kart board!

On a physical board capacity is best visually tracked on the board, with running totals under each person's or capability's name, and a running total to the side of the mix of what is on the board. You will likely round up or down to make these calculations easier. The rounding will wash out in the multiple iterations on the board. For example Dave might have 35/40 beside his name to show that 35 hours of work are on the board for him, leaving 5 that could flow on.

Remember, if something gets done a little quickly, a new iteration will just replace it sooner—and the reverse is also the case. We know from Grey Time that trying to get too accurate with estimates is just creating the illusion of accuracy, but not actually making anything more reliable.

The buffer board is your operations board—what everyone works to, and you will likely have another board for planning and management of the upcoming work—which we often call the 'bucket board,' very cleverly, because it is comprised of the buckets. Despite this, you will likely want to have any critical acceptability bands showing on the buffer board for the whole team to see frequently.

As an example of the number of iterations designed, ready for the team to work on, we might say 6 is far too few, 6 to 12 is a little too few, between 12 and 18 is great, 18 to 24 is a little too many, and 24 or more is far too many.

These can be displayed as a simple number, with a legend to show the acceptability like so:

Fig 23

The acceptability bands can also be shown more visually as some kind of gauge, which is especially easy on a whiteboard. Here are a few ideas:

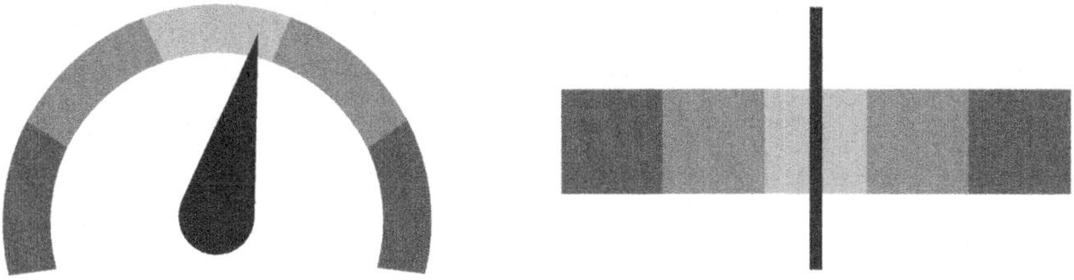

Fig 24

The more visual version is significantly more effective at getting external parties' attention, if they are the ones responsible for the acceptability band state. An image landing as an instant message or an email will tell them everything they need to know, including the implied nudge. A text-based message will be missed or ignored more easily.

Bucket 'Board'

Managing the buckets is far more flexible than managing the buffer. Even when you want the buffer to be physical for the benefits of adding visibility and tangibility, you will likely only treat some of the buckets as physical, and much of it will be digital. Physical bucket management can be a mix of the following:

Board—This is the same as a buffer, but flows through the buckets instead.

Trays—Good old fashioned 'in' and 'out'-style trays to keep track of all the work at its various stages.

All work in the system will need a paper note or iteration card of some sort to track it, as mentioned in the Schematic section. The point at which you detail the iteration card is the logical point to switch from digital to manual.

Bear in mind, that point must be at or before the final bucket, so the iterations can flow onto the board as soon as there is available capacity for them; if there is any prerequisite like writing out the ticket, it will get put off and the board will be an afterthought by the next day.

Physically, you will need either a large wooden board, a dedicated whiteboard, or something similar. If there are remote members of the team and you are using a physical board, you can host the Standups as a video chat, and send an updated picture of the board to the team at the end of each meeting.

CHAPTER 15

Standups

Updating the board

Live updates to the board are preferred for 3 reasons:

1. First, a live board means all the information is up to date and people can make relevant decisions at any time about what to work on. This, of course, increases the quick handovers of Pace and the smooth flow of work.
2. Secondly, it keeps the Standup meetings as short as possible, so people aren't wasting the whole team's time while updating details.
3. Finally, there is a satisfying feeling of 'ticking things off' that would be cruel to deny people.

In some situations, people will have missed things, or it will not be possible to keep the board updated live. For example, if some members of the team work remotely—or, in fact, if the whole team works remotely, and the board is managed by one individual. This person, or those in the office, should do what they can to keep the board as up-to-date as possible.

Standup meetings

Often referred to as stand-ups, these should be very tightly controlled meetings. If people want to discuss some aspect in detail, they should do so after the meeting; otherwise, the meeting will become a source of boredom and frustration. In that sense, it's no different than a daily Scrum meeting. I discussed in the Variation Management Rules that you can set your own rules; this is where they would be enforced. Similar to the Variation Management rules, you can also set your own questions for the stand-up. Here is our starting set we use when implementing Pace with a team:

Individual questions

1. Are you up to date?
   Purpose: Ensure you are talking to the accurate state (and make it uncomfortable to be unorganised[7])
2. What Zone are you in?
   a. If Zone 2 or Zone 3, move on to next person
      Let's keep the meeting short, when everything is on track it should take 2 or 3 minutes.
   b. If Zone 1: "Zone 1 rules apply, which are… (point at or list the rules if necessary)

   Remind everyone the Rules are non-negotiable (when enforced)
      i. What caused it?
         To buffer log, see Variation Management Rule 4
      ii. What's your plan?
         'Working harder' hasn't solved it so far and won't solve it now, we need clear actions to expedite and complete risky tasks.
      iii. Will it be cleared by the next Standup?
         We need to know the extent that the risky iteration will impact the whole team so we can take appropriate actions.
      iv. (Question to the others) Who can help?
         The information from the last two questions should help to assist with answering, both how a person can help, and if it is worth it (how at risk the iteration is).

7. For repeat offenders, keeping a visible tally of their repetitive out of date state will change this behaviour really quickly. No one likes their name on the naughty board in front of others. This can also be applied to people frequently late for meetings, keep a running total of their teammate's time they've wasted.

c. Optional question: What Zone will you be in at the next Standup?
Purpose: People will quickly increase their intuition of the movement between Zones by making predictions, and comparing the state at the next Standup.

Team questions

1. What Zone is the team in?
While an individual iteration can be managed by the people involved to get back on track, if a bulk of tasks are in Zone 1, it shows the team is not in control of the flow and far more drastic actions will need to be taken to get back in control.
    a. If Zone 2 or Zone 3, wrap up the meeting!
    Let's keep the meeting short—when everything is on track it should take 2 or 3 minutes.
    b. If Zone 1: "Zone 1 rules apply to all tasks, which are…" (point at or list the rules if necessary)
    Remind everyone the Rules are non-negotiable (when enforced)
    c. Optional question: What Zone will the team be in at the next Standup?
    Purpose: The team will quickly increase their intuition of the movement between Zones by making predictions, and comparing the state at the next Standup. This, in turn, makes them better at seeing the trend towards a team risk state and take steps to react earlier.

Management Visibility

Boards and visualisations can be mixed and matched in many ways to make a high-level, multi-team board. This can be helpful to give founders or execs an overview of the safety or risk of the flow of work through teams without all the detail. Like a series of pipes with pressure gauges on top, this multi-team board (affectionately called an Overlord Board by one founder) would likely show the acceptability bands for each team, and an indicator for the buffer position of each team. The relevant leader can then use this information as a source of positive reinforcement for the team and enquire where necessary. The aforementioned founder with the Overlord Board, represented on a screen in his office, would proudly show it off at conferences, showing others the amount of visibility he had over the flow of work in his business. If one team had too many amber lights, or a couple of reds, he could ping a message to them to see what needed to happen.

PART 4

DECISION-MAKING AT PACE

CHAPTER 16

Decision-Making in a Pace Environment

The human brain is incredible. It can learn and adapt to a multitude of experiences and skills. Humans have muscle memory, which is the term used to define a repeated pattern of physical movements, which make the movements feel more 'natural,' requiring less focus, and becoming more consistent. Along with muscle memory, we form a similar pattern in the connections of our brains, called 'heuristics.' Heuristics are rules of thumb that our brains form for decision-making, which make future decisions of similar nature much quicker and easier to make. This is an incredible feat of biology, allowing us to use past experiences to make present decisions, massively increasing the speed at which we are able to make those decisions. Such a feat, in fact, that it's worth stepping into this science lesson a little further.

Biology of Heuristics

How does this happen? Neurons in our brain contain cells, with dendrites extending outward, connected to branch-like axons. At the end of the axons are rootlike structures called axon terminals. These terminals connect to dendrites of other neurons within the brain, forming our neural networks.

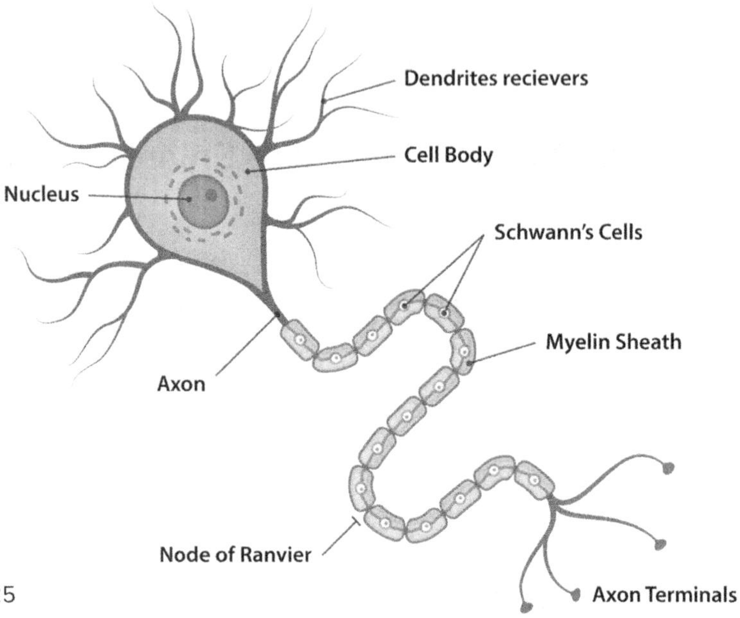

Fig 25

We learn new things via these connections, and we can also strengthen our understanding by adding more of these connections. Let's say you and I travel to modern Rome. If I have never had much interest in Ancient Rome, and you are a huge fan of the classics, we will retain understanding differently. If we go to a museum in modern Rome and take a guided tour, your brain will have a wealth of connections to then connect to further, from the books you've read and the podcasts you've listened to. I, on the other hand, have very few current connections about Rome to expand on, and will retain far less of the knowledge we learn on our tour of the museum.

This is the way we learn new things; but what about the way we increase our knowledge of things we already have connections for? Once the initial connections have been formed, they get stronger by having the connection insulated. The brain sends impulses along the axons to pass information and function. But not all axons are alike. The ones we use frequently become stronger and stronger, as our brain applies more insulation to them.

This insulation is a fatty tissue called myelin, and it accelerates the brain's signal along those connections that are insulated. A well-myelinated axon will transfer a signal 15 times faster than an un-myelinated axon.

Let's say you're learning French. As you practice and speak French more and more, your body will respond by wrapping more and more myelin around the axons involved, speeding up the signal. "Practice makes perfect" isn't just an adage, it's biological. This is how heuristics are formed; they are the same mechanical function biologically, just in relation to decision-making. This is really no different than deciding how to respond to a question in Italian, when the muscle memory of the same action happens on your nerve cells, as well as the neurons in your brain. This is similar to the way that learning to drive a manual car is tricky at first, but after a few years you can shift from 4th to 5th gear as you merge onto a highway without even being aware of it.

Kahneman's Systems

Psychologist Daniel Kahneman, famous for his work on the psychology of judgment and decision-making, identified the two primary decision-making systems in the human brain, which he named System 1 and System 2.

- System 1 is fast and relies on intuition. This system contains our last name, easy math equations, or how to change manual car gears easily three years after first learning to drive a stick. This includes anything to which we can respond without really 'thinking' about it.
- System 2 is slow, and includes our analytical mind.

For biological reasons, our body prefers to use energy-efficient processes, which System 2 is not. System 2 consumes significantly more energy; so much so, scientists can use a thermal imaging camera to tell how hard people are thinking by the heat given off, as metaphorical fans spin and our grey matter processor fires up. It would also be a huge hassle if we had to engage our analytical mind for every tiny decision we made.

System 1 is where our heuristics are based, and those heuristics are based on our past experiences. System 1 is also our default decision-making process, whereas we have to logically engage System 2. Even then, System 2 is biased by System 1, which can become a problem that we weren't even aware of. We might start making decisions based on what we've previously done, and not based on fundamentally good decision-making.

Practical Uses of the Systems

If you work a centuries old industry, and rely mostly on unchanging approaches, it's not really an issue. Software, however, is a fast growing, ever-changing environment. As such, we need to make good, robust decisions, fast, or we will continually miss out on opportunities, or even, fall behind the competition. On top of that if you're in a fast-growing company, internally you will need to make decisions without years of company experience to draw on. Let's apply this to a founder of a fast-growing software company.

When they first kicked the company off, it was just the founder and their mate coding at a dining room table. After some time, the company did well, and before they knew it they were hiring people. A couple years later, they are managing a team of people from a coworking space. The next step brings the founder to the terrifying leap of leading managers who are running their own teams as they make the move to a multi-team business. From there, it's not long until the founder is a CEO or manager of a medium-sized business, running multiple departments with teams ranging from development to sales with customer support, finance, and maybe consulting or projects in the middle. The founder finds themselves further extracted from operations as a figurehead of the company, speaking and drawing in new users at scale.

The wealth of experience this founder has built up is an incredible asset. But sometimes, they need to ignore elements of that past experience to make an objectively rational decision, and System 1 and all the myelin can get in the way of doing that.

We need to be able to make decisions from the fundamentals of good decision-making. Words like "rational, objective, logical, and cause and effect" get thrown around a lot in meetings, but without structure, how can you expect people to make decisions free from the assumptions that their world has installed in them? They need to employ practical structures of tools and processes which drive the fundamentals of decision-making.

Just like Systems 1 and 2 are neither friends nor enemies, myelin just 'is.' Using tools which support logical decision-making will myelinate those circuits. And, over time, logical decision-making will become your System 1. This isn't as easy as it sounds, since you're literally re-wiring your brain and then slowly laying down insulation on the new wires, but it's also not as scary as it seems. Lives are improved by better decisions—they're the core of everything we do—and great decisions become reinforced when we can see the outcomes of them. The tipping point of this system shift is closer than it seems.

The decision-making areas which most need attention are:

- Action focus
- Dilemmas
- Assumptions

These are most prevalent in meetings throughout software and many other industries. These areas are also where the most arguments happen, which lead to decisions made more slowly. Decisions are delayed as a result of multiple people with multiple heuristics attempting to make good decisions, and the arguments may be won out, not by the best decision-maker, but by the person with the loudest voice or the fanciest title in their email signature.

For an industry overflowing with smart and rational people who spend most of their lives writing commands in pure logic for a computer, this is pretty disappointing. In this section, we'll break down the four most common decision issues, how they appear, what to do about them, and how to utilize a practical tool (on paper, screen, or whiteboard) to resolve them—and build that myelin on the important axons.

CHAPTER 17

Application to Decision Making at Pace

The way the above techniques and skills can benefit an individual is probably self-explanatory, but how do they benefit teams and the company as a whole?

Accelerating the speed at which a team can produce code and the quality at which they can produce it is one critical half of a software team's effectiveness and the scaling of a software business. The other half is applying those improvements to building the right things in the first place. It's more efficient to do that than to simply go faster in the wrong direction. For this to happen you need two particular elements to be true: you need teams that make autonomous decisions, and you need business-level decisions being made that truly solve your customers' problems.

1. Autonomous decision-making of teams

The more teams can make good decisions on their own, the better your business can scale, and the more effective teams can be. This also means leaders can spend more time each day focusing on the growth and future of the business, rather than putting out fires. You can test autonomous decision-making by considering how involved people outside the team are in their decisions and their direction. People external to the team may be pulled in because the team is not confident enough to make their own decisions and therefore delegates the decisions upwards, or they may intrude because managers don't trust the team and force themselves in to make decisions for them.

2. Creating valuable solutions for customers in your software

You will immediately recognise that this is really an extension of the first point, as the teams are where you want these decisions to be made.

The skills and techniques required to create these solutions are also extensions of those same techniques used for autonomous decision-making. The difference is applying them to your customer's heuristics rather than your own—which, as you might suspect, is much more difficult. To test this, determine whether you deliver features that are requested by customers, or whether you use customers' requests and feedback to identify and build functionality that gives them what they really wanted, or even more than what they originally wanted.

Action Focus

We often get caught up in the specifics of our actions, rather than examining the effect or purpose of the actions in the first place. This can be exacerbated by the implied need in many businesses to make quick decisions, the expectation to react and solve a problem. If a current activity is no longer viable, there is an amount of pressure to quickly come up with a replacement action. There is rarely pressure to take this as an opportunity to consider an even more effective alternative, and to take the time to really think through a better action for the future. And even when we do, if we don't have the skills or techniques to do this, it can just feel like a hopeless activity.

To open up to alternative ideas, we need to instead consider the effect or purpose of any action we're taking. By focusing on the action, we look to replace only the necessary elements. For example, let's say you have a monthly Standup meeting with some of your development partners or beta clients, perhaps for a new module you're working on or some part of an implementation project. Ideally, you meet with the project champion or team leader from the client site each month to gather and discuss user feedback in the central place. Then, imagine you encounter a situation where this can't happen—perhaps you take on a beta client in another country, or the project champion moves to a different location to work from— and you can no longer have your monthly in-person meetings. You might intend to replace one action (the in-person meeting) with another similar action (a virtual meeting over Zoom or Teams), as this seems sensible and logical; however, this would be ignoring the purpose of the original action (the client Standup meeting).

Instead of replacing this action, you should instead uncover the reason it was implemented in the first place. What was the purpose of it? Instead of just focusing on how to virtually replace the action, you should uncover the purpose of the action to open up more effective options for achieving the purpose.

In this example of the client Standup meeting, the purpose is to get user feedback from the individuals—in which case replacing it with a virtual meeting might not be the best way to proceed. Is a virtual meeting really the best way to get user feedback? Instead, what might be more useful is a direct channel to the actual users via a Teams or Slack channel, giving everyone a way to shoot through feedback in real time, and to communicate directly. Perhaps there is yet another idea that works better; there are multiple ways for users to give more direct feedback. What this process does is open up options for getting the right effects, rather than focus on how to replace one action with another.

This gives you the opportunity to actually make better decisions built from a more effective analysis of the situation. You might be reading this, thinking, *Well, a meeting like that isn't about the user feedback for us, we get plenty of updates from users. Really, for us it's about using the face-to-face interactions to build rapport with our contact at the client end.* In that case you have a different purpose, but the technique of considering the purpose doesn't change; instead, you can just consider other ways you could achieve the effect of more rapport with customers. Which, again, opens the door to many more options than simply considering how to replace the in-person meeting.

Maybe you can personalize the updates that you send out by including details from your team's personal lives, so as to promote familiarity, which in turn builds rapport. If your business is built on the value of genuine relationships with your customers—caring about them as users and people—this will create successful rapport and personal connection. If your business wants to stay in touch and keep socially connected—so your company doesn't begin to feel too 'corporate' or stale—perhaps a better option would be to host a virtual quiz night with friendly competition between customers.

Now, remember, we don't want to simply look at the actions, so I'm not saying that those specific examples are things you should be doing. Rather, the important method here is to stop focusing on the actions, and to focus instead on the effect you're looking to achieve from the action.

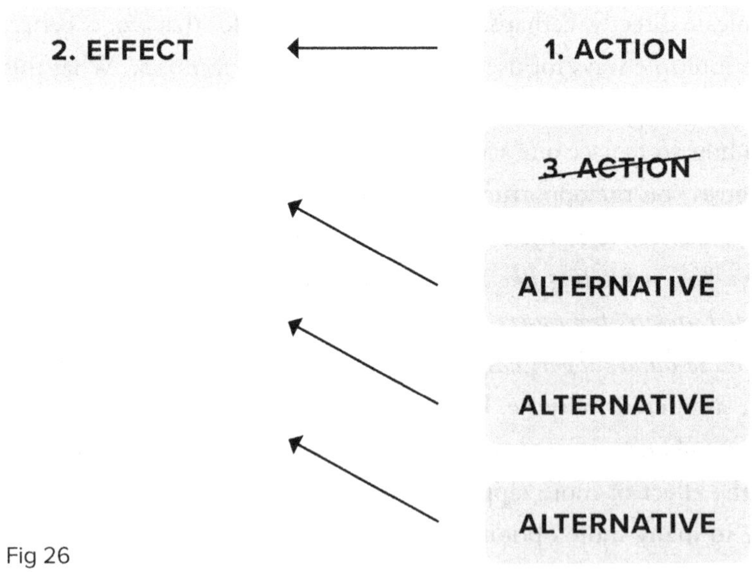

Fig 26

Assumptions

Now that we've explored a couple of quick daily techniques, it's time to dive into the skill that has the potential to literally change your life. Mastering this skill changes the way you view problems, makes you a clearer thinker who is less influenced by your heuristics or the noise around the information, and makes you a more effective decision maker. The value here is immense, and I want to make certain it is not overlooked. I can confidently say this is the most valuable part of the book—it opens System 1 to System 2 in a way no other technique I've ever seen can.

Heuristics and System 1 are compiled from underlying beliefs or assumptions about the way things are or the way they must be. These assumptions are also hidden in our System 2 thinking—that is, when we think logically in System 2, our logic is limited by these beliefs, which our brain regards as rules or parameters for thinking. If we had none of these heuristics, we would have to think about literally everything related to our topic and never be able to reach a decision. Carl Sagan's words apply to this thinking: "If you wish to make an apple pie from scratch, you must first invent the universe." Because we have seen gravity play out many times in our lives, we can safely conclude that if we let go of something it will fall; our heuristics are at play here, assuming all these parameters, so we don't have to consciously think about them." Because we don't have to think about them consciously, other assumptions that are less robust than gravity will influence our decisions. For example, "If I tell Greg about this, he'll yell at me" might be something from past experiences that means we are only thinking about how to solve a problem without talking about it. There are an immeasurable number of these assumptions being made, all with varying degrees of validity and relevance. So, what we need to do is go deeper into the usage of System 2 and turn it to both identifying these assumptions, and challenging their validity.

Any time a change is forced upon us and we are forced to make a change which we previously didn't agree with or didn't believe we could make, we are challenging these assumptions subconsciously. The COVID-19 pandemic of 2020 has made this abundantly clear. People who had a laundry list of reasons of why they couldn't work from home, or reasons they absolutely had to travel to visit customers to do their jobs effectively, are now doing 'the impossible' and working from home. Some of those people and companies will find that they never go back to their old ways, while others are making do for the moment and will resume their old ways when they can. The perspective has been forced upon them like never before; they've actually had to consider it for the first time, and I strongly believe many of those who go back to their old ways will do so because they never tried to think seriously about how to be more effective in the changed situation, and instead focused on how to merely get by. I am also confident that this sticking to the status-quo will be proven the wrong approach by the companies who do embrace a different way of operating for the better. This will be the global version of Kodak, which—despite inventing digital photography—did little to move towards it, preferring to invest in better film technologies. Ultimately, Kodak declined to bankruptcy (ironic for a company that was hugely successful by being a first adopter of colour film despite many others focusing on black and white due to the higher [at the time] quality).

There are also assumptions held at team, company, national, and industry levels which are opening huge windows of opportunities for those who can identify the faulty assumptions and provide a solution that challenges them. I'll cover this in the section below on assumptions for others. First, we need to become skilled at identifying our own assumptions and challenging them. To do that effectively, we need to be able to define the situation by using logical structures. There are two of these: Necessity and Sufficiency (or 'Pre-requisite' and 'Cause and Effect' in more common language).

- Necessity (or Pre-requisite) is where something is required to be true before we can have something else, i.e. "In order to have X, we must have Y." In everyday use, we can use this more casually, saying, "In order to prepare a coffee, I must have boiling water."
- Sufficiency (or Cause and Effect) is where something (or a group of things) is sufficient to cause an outcome or effect, i.e. "If X and Y, then Z." In everyday use, we can use this more casually, saying, "If my car is working and has fuel in the tank, then it will start."

You probably noticed that both of these statements are full of assumptions. You could quickly raise an assumption that, if it wasn't true, would make either statement invalid—this is important to the process. In a scientific theory, we would be far more robust about our statements and details of the sufficiency or necessity logic. However, the statements above are just fine for this purpose, because we are looking to find the assumptions we are making but haven't stated. This technique can be applied to either sufficiency or necessity; however, it's easier to do with necessity so we will start there.

Looking at the coffee example, we can list the assumptions we're making with "In order to prepare a coffee, I must have boiling water" because:

- Coffee must be boiling when prepared.
- It is not coffee if it's not hot.
- Coffee cannot be prepared without water.

This is just a short list of assumptions; however, the range in the list should show how many different assumptions we're making about the way things 'have to' or 'should' be. Many of the assumptions we identify will be valid—or, while challengeable, we would prefer they remain true. However, there will be some that are challengeable that we wish to change to solve our situation. As you can see from the list, only one of those assumptions needs to change in order to alter the necessity requirement of "In order to prepare coffee, I must have boiling water."

How to use it in software:

With the basis of identifying and challenging assumptions explained, let's look at how that can help build autonomous teams. In many cases, people will claim they didn't have the information or didn't have the experience when they have made a poor decision. There are cases where this is true, and the technique here is by no means a magic provider for perfect decision making; however, in many cases people have enough information and experience to have made much better decisions or to solve problems far more effectively, but they haven't applied that information in such a way as to achieve that.

With the assumptions technique they can apply that information in a more effect way. As we know from Myelin and System 1, the more they do that, the more it will become a subconscious skill. Teams and decision makers need to identify the assumptions under the decisions they are making. Let's take the ever-present situation of trying to improve the speed of development without compromising quality, or the quality without slowing down. (Obviously, by reading about Pace in this book, you now have an unfair advantage!)

A team leader asked to improve speed without compromising quality might get really frustrated with management, or they might compile a list of reasons why this can't be done, or they might push this task upwards saying they don't know how to achieve it, or they could come up with a few small changes that don't amount to much, or they could make some rash decisions with serious negative effects just to get on with it. Alternatively, let's say that before coming up with a list of solution ideas, they first consider what assumptions they're making about how they develop software now. Since we need a statement to start us off on assumption raising, we'll say, "In order to complete functionality faster, we must compromise quality," because:

1. Faster development always produces bugs.
2. The team forgets things when rushing.
3. Short-cuts are taken when reviews are done quickly.
4. We can't go any faster with the people we have.
5. The people we have don't have the capacity to do more.
6. Quality steps slow us down.
7. Our technical debt rises the faster we go.
8. Unqualified people approve things when we're under the pump.
9. Our team pays less attention to the specs when they're working fast.
10. Quality considerations are the first to go with speed.
11. In our team, it's either on-time or on-spec, not both.
12. We've never introduced quality processes that speed us up.
13. People's primary review focus becomes, "Is it good enough?" not, "Can it be better?"
14. We use speed as an excuse for not doing a good job.
15. Rework is quicker than getting it right the first time.
16. More frequent test steps make the functionality take longer.
17. The more small reviews we do, the slower we go.
18. The user gets less input if we're going faster.
19. UX becomes a 'nice to have' when we have a lot of work on.
20. Writing test first takes away from valuable coding time.
21. Devs can either spend time coding or testing, not both.
22. We can develop faster to meet metrics now, because bugs show up later or not at all.

Given the context of a specific team, and time, the list can become very large. In reading that, there will be some assumptions on there that you will agree are true, some which you think could be challenged, and some which you think, *Yeah, that one looks easier to challenge*. Let's take an easy one, "Rework is quicker than getting it right the first time".

Remember we are trying to pick holes in the opinion that speed comes at a cost of quality. So, if we read the whole statement out with our chosen assumption we get: "In order to complete functionality faster, we must compromise quality because... rework is quicker than getting it right the first time" When read like this, the statement just doesn't seem true. Instinctively, the problem solvers among us will be triggered to come up with ideas on how to prove that rework actually slows down the speed of development. This opens up a whole new conversation on how taking the time to slow down while coding, i.e. reducing rework, will in fact speed us up while preserving quality.

Great—we have an idea, now you can brainstorm ways to implement it. These ideas might include: Put a review 20% into the dev work to make sure it's on track, or ensure a test first is written, or many other possibilities including those mentioned earlier in this book! The value is not in this individual outcome (although it is of value); the primary value is in using our System 2 brain to properly analyse and challenge the status quo to lead to genuine breakthroughs. The secondary effect of this is that, as our team leaders use this technique more and more, they increase their System 1 skill of subconsciously looking of chinks in the armour of problems and quickly seeing the assumptions they and others are making.

Dilemmas

Making decisions in isolation is not the only time we need to take the focus off of actions. We also need to do it when faced with a dilemma between two opposing actions we can take. Our first instinct is to compare the two actions, sometimes spending hours focused on them or agonising over them in the back of our mind. We can apply the same logic from the first technique here as well, to identify either a way to break the deadlock or to come up with an entirely new option, which will give us even more benefit.

To achieve this, we consider the effects we want from each of the actions. In this way, we see that even though the actions may seem mutually exclusive, the effects we desire from each of them are not mutually exclusive. Then we look for a third action that is totally different from the first two, but that will either give us the effect we want, or will give us one of the effects while not conflicting with the other action. Alternatively, sometimes once we understand the effects of each action, we realise one effect outweighs the other so significantly that we just choose the action that serves it, and get on with it. We'll look at more detail on how to do this in the 'With Others' part below.

CHAPTER 18

Decisions Involving Other People

When using these techniques to involve others, there are really two key ways they can be applied. It can be collaboratively, where we use them as a team or with another person when debating a decision. Or they can be applied on behalf of others to benefit them, such as when considering the best way to develop new functionality to benefit our customers and users. The two situations are somewhat different, with the first being one where you seek to both get clarity on the situation and work together on the output. The second situation involves looking at what the users are doing, and what the customers are asking for, and identifying why they are doing and asking for those things, so you can deliver something which delivers higher value.

With Others

More often than not, when dealing with others, will be using the first couple of techniques I introduced, the effects from assumptions and dilemmas. This is because in most cases like this, you (or you and a group) will be on one side of the dilemma while the other person (or group) is on the other. You must first clarify the dilemma; as we discussed earlier, situations like this will start with both parties looking at the actions to be taken or the mechanisms involved, rather than looking at the purpose for those actions in the first place. Your first step will be to define the two options that are being discussed, then to define the effect that each of the actions seeks to deliver. If you draw it out, it can look something like this:

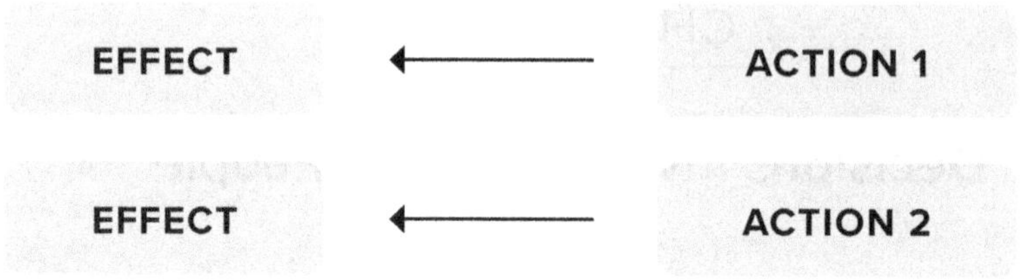

Fig 27

This will often bring both parties together to focus their efforts on solving the problem, rather than just arguing with each other. If the common goal is still not clear for the group, you can add a fifth statement in a column to the left of the two effects which defines the common goal the two effects are seeking to deliver. At this point the diagram becomes an 'Evaporating Cloud', a tool in a toolset called the 'Thinking Processes' designed by Eli Goldratt.

Once you have the effects clear, you can follow the same steps we went through earlier, in order to decide whether there is another possible action—one we can take that will result in one of these effects which doesn't conflict with the action for the other effect. Or, even better, is there an entirely new action we can take which gives us the benefit of both effects? To really get to some powerful solutions, you can add in the assumption technique we looked at too. The way we have been stating the action and effect coupling is necessity: "In order to get the effect, I must have the action." This means you can follow the assumption identification steps and challenge the weaker assumptions to find a solution.

You will find that, in some cases, you don't even need to get this far; simply clarifying the effects of your desired actions to each other will cause one of you to say, "Oh, THAT's why you're doing it, okay, that's easy…I'll just do this other thing instead." To get to this outcome quickly, you can use the effect focus before you even get to the dilemma stage. When someone proposes an action you don't understand or agree with, simply ask their desired effect from the action. The inverse of this is if someone else reacts to an action you propose or are already doing, before defending the action, tell the other person your desired effect that is driving the action. Both these approaches will resolve so many conflicts before they occur, and will cause far more useful discussions that are focused on the effects or benefits of things, rather than the actions and mechanisms used.

An example:

Let's say Dave and Ritika from Xyphyr are arguing over a feature in the software—a classic case of complexity vs. usability. They want to add functionality for reports for managers, and Dave is proposing they make the functionality really simple—something they can access via their laptops or their phones or tablets on the go. Ritika, on the other hand, thinks that isn't enough, and she thinks they should take the time to build out the functionality more to have more powerful configurations. Dave and Ritika are both aware the team has plenty to work on and can't commit to a complex configuration tool with enough time put into the interface to make it really usable, too. They've only been discussing for a few minutes, but their volume is increasing as they each try to stress their own point.

Chris walks past with his coffee, hearing the voices rise, and stops for a few seconds. Chris watches with intrigue, knowing he's been dragged in to arbitrate such discussions far too many times, and far too many more times been dragged in to 'make a ruling.'

Dave and Ritika pause their wild gesticulating and look at him. Just as Dave opens his mouth, Chris says, "Why don't you try talking about effects instead, like we read in the book? The technique is listed in steps in there," pointing at the copy of *Pace* on Ritika's desk. (Cheers for the plug, Chris.)

Dave picks up the stylus for his tablet, while Ritika grabs the book off her desk. Dave writes 'simple functionality' in the top right corner of his tablet screen, and 'complex functionality' in the bottom right. Ritika sits down with the book in her hand as Dave mutters out loud, tapping the top left of his screen where the effect will go.

"Well, I said it should be easily usable, but that's not the effect, really. I think the effect I want is that they use it a lot, that's why we want things to be usable. I want managers whipping out a quick report to see the answer to the question they're pondering as quickly as they would Google something," Dave says as he writes, "Managers turn to our reports first for information".

"Oh, I didn't think of that; that's actually a good point," says Ritika.

Dave carries on, "In order for managers to turn to our reports first, they have to be simple; yes, it assumes simple is usable, but that's generally true for us."

"Yeah, okay, so for complex, I guess really I want managers to always find the answer in our reports," says Ritika. "So, does this make sense: 'to have managers always find the answer in our reports, we need to have complex functionality'?"

"Yeah, it does, and we can't lose that," says Dave. "I'm assuming the complexity will increase the chance of finding what they want—are you also assuming they'll configure it themselves?"

"Oh, no, it will be a techie minion of theirs, no doubt," says Ritika.

"In that case, do they even need the configuration functions, or can we just give them a way to dump the data?"

"Actually, a data dump is even better," says Ritika. "I've literally heard you yell at your screen, 'just give me the bloody raw data!' more than once."

"Yeah, true," laughs Dave. "Okay, so we'll build the simple and quick reports, and give them a way to just dump out the data they want, where someone can put it into Excel or Power BI and do what they want with it? Perfect."

"There you go, you can't just read about robust decision making, you have to actually use the techniques to get the benefits," says Chris smugly as he strolls off.

On Behalf of Others

Looking at the assumptions others are making is a skill which is rather difficult, but very powerful, and one shared by the world innovators. There are two main settings this can be applied to:

1. Understanding the assumptions of your end users and customers.
2. Applying it to your target market as a whole, identifying and challenging the assumptions that are held as facts or 'the way it has to be.'

Both of these applications will require you to have built some decent skill in the technique itself—so before you go trying to understand others, you'll need to have fluency in understanding yourself to avoid making assumption errors. With that in mind, let's look at understanding the assumptions of your end users and customers in more detail.

Customers

Customers and users will often ask for things that are not the ideal way to get the effect they want. It is very likely they haven't considered the effect, and it is also true that people don't know what they don't know. That means the likelihood that the action (often a feature) they have requested is not the best way to deliver the outcome they want. On top of this, we must ensure we don't forget there is a difference between a user and a customer for many software companies. For many, the person or people choosing to use your software and paying the bills are not the same people as those who are the end users. We must make sure we don't develop features for the user that don't deliver value to the customer. When looking at customer requests, we want to consider the effect of what they're asking for, as mentioned above.

A classic example of this—and this is not an endorsement of this approach, but it makes the point—is a software suite designed for small gym owners. We will call it FitApp here. FitApp is by far the dominant player in their niche, and users absolutely hate the apps. The reviews are among the lowest and most scathing you will see in the App Store—they are buggy, unstable, and all around a poor user experience. So how are they so successful? The gym owners are the ones paying the bills, and gym owners generally start their gym to do what they love: training others and helping others reach their potential. What they don't love is all the admin and digital infrastructure of running a gym. FitApp takes all of that away—handling contracts, waivers, payments, class timetables, bookings, special events, and more! This is why the gym owners tolerate the negatives, and make their members use the app. One end user benefit is the ecosystem; if you attend another gym with the same software, you are already a user so it is easy to connect without being required to sign up again. With all that in mind, they have done very well. Once again, I'm not stating this as a desired situation to target, by any means.

The point is, the FitApp's owners understand the problems and frustrations their customers have better than most of these customers could articulate. Many of these frustrations are rooted in the way the customers have previously assumed things needed to operate. The gym owners likely never considered the alternative until FitApp company came along. There are incumbent players in the fitness space who could have captured this niche too; however, they did not. They likely operated with the assumptions they had from their current target market, and failed to understand the different assumptions of this gym owner niche.

To understand what would really provide value, follow the steps outlined earlier. Start by seeking to understand the effect of anything your customers request—for some, this will be as simple as asking what they want from that feature, or how it will benefit them or their team. Others may not know why they asked, so you should do some digging to get a feel for the reasoning behind their request. Once you have clarified the effect, ideally from more than one user or more than one customer, use the assumptions process to logically understand what they're assuming.

Look for the missing understanding in their request so you can improve the outcome. Do this exactly the same way as described earlier. If the customer says, "In order to increase our client's spending we must be able to customise every email template," identify the assumptions. You can check some key assumptions with them; if they are correct, challenge them and find a way to implement the change.

Target Market

If customer application seems difficult, applying this to a target market is incrementally more difficult—but the potential reward also increases in magnitude. When looking at your whole target market or an entire industry, you will not be focusing on individual features and requests. Instead, look at what fundamentally is making it difficult for them to achieve their goals. If this is a company, what makes it hard to be this particular company? If it is an industry, what problems does the industry have, that are so common and entrenched as accepted truths or requirements that people in the industry are totally blind to the underlying assumptions?

Henry Ford said, "If I asked my customers what they wanted, they would have said a faster horse." This was re-quoted by Steve Jobs who, among many innovations that challenged the status quo, created the iPad—a tablet which doesn't require a stylus. Tablets had previously existed with limited success. Look at the early touchscreen Windows phones (which literally had a tiny desktop, including a start button you needed to stab with the stylus) to see why. One of Jobs' requirements was touchscreen devices which people can use naturally without the obstacle of a stylus. This was something 99% of users had certainly not considered.

Some may have attempted in the past to change the way tablets were used, but failed—which just reinforced the 'truth' of the matter to others. Space X slashing the price of launching things into space with re-usable rockets is an example of another company challenging the 'truth' of the way the space industry has always been. Considering these examples, it's clear to see that people within an industry can only provide some of the information you need, and the rest will come from watching, learning and uncovering assumptions.

Look at what people complain about, look at why businesses in the industry struggle or fold, look at what those who stand out do differently. This will require some seriously robust analysis; as you can imagine this is a strategic level analysis, not something a team would knock off in a few days. By building solutions that solve fundamental problems customers have in an industry into your software, you will secure your software as an extremely high value service which allows you to charge accordingly (which can be magnitudes more than you can charge as a commodity provider). Thus you will set yourself so significantly apart from your current (or would-be) competitors into a market of your own where you have few competitors.

Wrap-Up

This whole section is about logical analysis and provides some tools to assist. But we don't need to limit ourselves to these. Look for any tools that help with this. A common example is the '5 Whys'[8]: dig into the cause and effect by working backwards with the 5 Why's. Once you have the chain of cause and effect, you can look into the assumptions of any of the connections. Find tools that work for you and your team; but most importantly, treat thinking in the same way you do working or processes. Every company looks to actively improve the way people *do* what they do, and improve the way the company manages what people do. Rarely do they treat the way people *think* about what they do with the same regard.

8. I won't explain the 5 Why's here, but you can easily find info online about it. 'Essentially ask why x happened to find a cause, repeat 5 times or to a useful level to get the full picture.

PART 5

IMPLEMENTATION TIPS AND TRICKS

CHAPTER 19

Likes/dislikes

All that's left is to discuss how to actually implement the changes we've discussed above. Ultimately, the success of the implementation process is made up entirely of human factors. In flight training we learn that 75% of accidents are caused by pilot error. I would hazard a guess, based on all the projects I've been involved with, that it is the same rate for change project failure. This means that 25% of the time, the changes and configurations themselves are a problem, but 75% of the time the problem is the people who are not taking the right actions and instead falling back on old habits. In this section I have provided a few of the key tools we use in implementations to make immediate changes—and cause them to stick.

Overview

One of the first steps of a project is to talk to everyone involved. If the business is small (under, say, 50-ish people) we try to speak to everyone[9], regardless of their involvement in the project. We speak to everyone for a few reasons. First, we just want to break the ice—we want people to be comfortable with us, and the best way to do that is to have a brief initial conversation that builds a bit of rapport. It does wonders for lunch-room-style conversations too, where staff will fill you in on some 'informal'

9. I mean everyone, from the receptionist to the CEO. The danger of missing even a couple of individuals is those people feel left out, and become obstacles!
An example was the maintenance guy (a part-timer we didn't speak to initially), who saw us talking to everyone and was very grumpy every time I passed him by. One of the managers even mentioned he is a bit edgy we were there. Then once we did Likes/Dislikes with him, he had a ton of stuff to say, and was extremely friendly afterwards.

truths that they don't think the managers will share. Second, we want to know what is working for people and what isn't—a great way to quickly get people on board with a new solution is to make it more beneficial for them than the current situation. Although we are making many changes, we will try to keep many of the things that people like, and remove many of the things that they don't. As you have seen, there are critical components of Pace that are non-negotiable; however, you have also seen that it is a very flexible solution in many other areas. In the flexible areas we can adjust things based on the likes and dislikes of people. Finally, getting to know people as individuals rather than just workers means that you can have positive conversations with them about things they're interested in.

I remember a team member at a company in Germany who was a fan of the Philadelphia Eagles (a professional American football team) because her son loves them. As you can imagine, no other German in the office was particularly interested in American football. So, you can imagine what it meant when, a few weeks later on the project, I congratulated her on the team's Super Bowl win! This informal chat can create a personal connection and great rapport.

The primary risk in this step is that people often think they are on trial, like a performance or redundancy conversation. There isn't much that gets people more on edge than hearing, "Here are the consultants we've hired to look at our processes and increase efficiencies." People only hear, "We're trying to find people to get rid of". Interestingly, we would never describe what we do that way—as it isn't accurate—but others often make this assumption on our behalf. Creating these personal conversations means that we are approaching the people, whether that means popping over to their desk, giving them a quick call, or a video chat. Initially people find it strange and awkward, but once you've spoken to a few people the word spreads, and you can tell that people are excited to say their bit once the assumed fear barrier is broken. Make sure you don't bring a notebook to these conversations, as that raises a red flag that the things said will be used against them at some point. This breaks the banter, making it more of a formal process; jot down things later if you need to.

The Process

Have a chat with everyone involved with the project, and everyone in the company if possible.

Ask these questions, but engage in a discussion with them—this isn't an interview, it's a chat (you have to share things, too).

- What do you do here?
- What do you like about what you do?
- What don't you like about what you do?
- What do you like to do when you're not here?

The first will break the ice, the second and third provide valuable information for changes, and the fourth lets you get to know what their interests are outside of work.

CHAPTER 20

Setting the goal to track productivity

Assuming that the fact that you're reading this means you're in the software industry, you won't need to be told that measuring productivity output in software could be considered rather difficult . . . or a bloody nightmare. There are hundreds of different approaches by various companies all over the world and they all have their own pros and cons. You may have an approach you prefer, or you may have, more or less, given up. Having visibility of the Wins is huge to maintaining momentum in a change project. The productivity gains are easily lost amongst all the activity if they aren't focused on something in particular. Many teams do multiple types of work, so you should choose a single type to measure. For example, if the team works on bug fixes, customer requests, and new features, choose one of these to focus increases on as productivity rises. As you implement the phases of Pace, you will increase the productivity, so only schedule in extra work of the chosen type. To get the baseline, review the last three months of completed work. Take an average of the iterations or check-ins of each—for example 10 bug fixes, 5 customer requests, or 2 new features each month (some types of work will be longer to complete per iteration than others, we want to use actuals instead of percentages).

From this baseline, choose one type of work you want to increase—for example, bug fixes. Now, as the team gets more productive, you should seek to maintain 5 customer requests, and 2 new features per month while watching the number of bug fix completions rise. This manages the 'noise' by giving a measurable control point.

Which type of work you select, and the length of time you maintain this are very important decisions. You probably already know which type of work is most limiting your team or business from scaling to your goals. If not, I'll give some prompts. If you feel like the teams are starting to drown in bugs and new developments are starting to crawl, you might even be getting close to the bug vortex—where the new bug rate is so high there is risk it will grow to the point where it suffocates the business. In this case, the choice should be clear. Perhaps it isn't bugs but cashflow that is the struggle right now, and a few more of those paid customer requests or customisations would make a world of difference. Or your marketing and sales teams might be fighting a losing battle with the dated feature set you put forward against your competitors. The choice here is not only critical to measuring success, it can be critical to the business's success (or survival). For this reason it is also critical to maintain support of the people paying the bills (if that isn't you) whether that be the owner, the board, the finance team, or external investors. They're going to want to see these data to back your decision and get excited about rolling Pace out to more teams!

CHAPTER 21

Wins board

We touched on the importance of tracking Wins in a tangible way to measure productivity and outcomes above. What about the intangibles? We need those to be visible too, both for the people involved and for the external parties mentioned above to see the human benefits. Create a Wins board right at the start of the project, a central place for anyone to write up the wins they have experienced as a direct result of this change project! This should not be curated, but should be encouraged. Some things that are wins to some, won't matter to others, but it's the collection that matters. One person's win might be the fact that they wrapped to go home on time every night to see their kids; another person who isn't bothered by finishing time is just happy to have the visibility necessary to not have to annoy people throughout the day, asking for updates. The things that go on the Wins board will surprise you, and its effect on the team will surprise you too. These are little things we can miss or quickly accept as normal once they change, but they can remind us of how far we have come!

Process

Find a visible place, either digital or physical, to set up the Wins board. Advise everyone of the purpose, then listen for small wins as you start making changes. As you hear them, encourage those people to add them to the board to get it started; after a couple of encouraged updates, people will take over and add things without prompting[10].

10. During a Pace project in Ireland, we give out sweets to people as they wrote stuff up. We gradually transitioned to just leaving a box-full of sweets next to the wins board so people can help themselves as they wrote up wins. Worked a like charm!

CHAPTER 22

Go before you know/Know before you go

Change projects that include a paradigm shift are more difficult to implement than those which have incremental changes. We discussed in the section on autonomous decision-making that this is because people understand things and make decisions based on their previous experiences, which build up their system. Paradigm shifts, by nature, are things to which people cannot apply their experience or intuition; what makes that worse is that when technically-minded people don't understand something, they try to get more detail to understand it more deeply. In this situation, more information doesn't help because people are mostly checking it (subconsciously) against assumptions that are not valid. They have spent so long building and reinforcing assumptions around the previous paradigm that, the more information they get, the greater number of things they have that don't match their intuition, and are therefore less and less likely to agree with the changes.

There are two primary ways to handle this, and we would recommend both. The first approach is the 'safe harbour' approach—this involves moving from a stable state to the next incremental stable state, then assessing before implementing the next step. A simple way to break a full Pace install down is to break it into the three sections we reviewed: Bottleneck Management, Variation Management, and Flow Management. By following these you are mostly building on people's experiences at each step. You will find Flow Management the biggest step if you currently run sprints.

The second approach is called 'Go before you know.' This involves understanding the solution in detail before implementation, or implementing and then using the experience to understand the solution. The answer is, of course, a balance, but if people are trying to understand the concepts and getting stuck in the intuition trap mentioned above, you should move to a 'let's try it and see how it works' approach. Sometimes this will just involve trying it; for example, adding a buffer to your sprints isn't going to bring the company down if it doesn't work well on the first sprint. Other times you may wish to do a trial somewhere in a safer environment; for example, implementing a relevant mix of tasks (Flow Management Rule 4) would be more helpful in a customer service team or team with tight service level agreements than it would be in a team making new features for a module that is not yet released.

CHAPTER 23

I'll do it on something important vs. unimportant

This is a similar predicament that people can find themselves in when learning and potentially applying new tools and skills. There are a number of reasons people do this to themselves—from a fear of 'wasting time' to not wanting to be wrong or look silly in front of others. The dilemma is that people choose not to apply a tool or change to something that doesn't really need it, in order to practise. However, when they get to something that actually needs changing, they haven't tested the tool or change enough to be confident in trying it there, either. So it sits as a nice idea in theory and never gets used.

The only way to resolve this is by compelling people to try out the new idea on menial things and to measure the number of attempts, not the quality of the output of those attempts. For example, in learning a new decision-making tool, people should test it on decisions like, 'should I have tea or coffee?' or, 'what shall I have for lunch?' Even if they know what they will pick, it builds skills with the tool and helps them to understand the mechanics of it in such a way that they can apply it in more important circumstances. The same applies to changes and change projects; often the statement goes, 'oh, this team is going alright, they don't really need the change,' and, 'oh, that team really needs a change but it's not wise to start there!'

Sales trainers know this from getting trainees to just pick up the phone and use the script on anyone, rather than saving the 'good script' for a really good opportunity.

CHAPTER 24

One-Liners

Cultural change is one of the most difficult parts of any change project. It's a lovely buzzword, but what do I mean by that? Cultural change refers to the group habits. Habits are something individuals do without thinking about it (here we are looking at System 1 again!); the team or organisational culture is made up of the habits that apply to the majority of the individuals. They are particularly strong because, not only does the individual have to overcome the System 1 driver, they also need to overcome any social pressure from those who aren't winning against their System 1 driver.

To illustrate the power of this nonverbal social pressure, a great experiment was conducted in a documentary on group think. The experiment was conducted in a stereotypical waiting room. The first experiment participant enters (not knowing they are a participant), signs in at the reception, and sits down amongst the others (who are all actors). After a few minutes, a loud tone sounds 'BEEEEP,' and all the actors calmly stand up and then sit back down. They do this while reading the paper or playing on their phones; meanwhile, the first participant just looks up, confused. The second time the loud tone sounds, the participant sheepishly raises herself slightly out of the chair. The third time the tone sounds, she slowly stands, and by the fourth beep she's fully standing each time. At this point, those running the experiment begin to gradually send the actors out of the waiting room and into the rooms beyond, and usher in new participants. Of course, the first participant is now doing what the actors were, so the new participants also follow her lead and join in, standing and sitting each time the loud tone sounds. By the end of the experiment there are no actors, just a waiting room filled with participants who are all standing up and sitting down to the tone; none of them know why they are doing this, but they all feel pressured to continue.

To overcome the cultural habits holding back change, we need something that our first movers (who are able to overcome their own System 1) can do to snap the others out of the habit. A surprisingly effective and simple technique here is the One-Liner. A one-liner is a 'saying' people can verbalize that reminds others of the new habit. These can be used in a meeting, or during casual discussions.

Here's a couple of examples we use:

One-liner: "Shoving it in doesn't make it come out."

Meaning: Without massively compromising quality, the team is going to get done what it can get done for its capacity. Anyone 'shoving work into the team' (meaning anyone attempting to add unplanned work in the middle of a plan, e.g. a sprint) should be told no, or if rejection is not possible, that the unplanned addition requires something of equal estimate/capability to be removed from the current sprint.

When to use: Either coyly to the person attempting to do the shoving, or to remind everyone during a team discussion.

One-liner: "What's the effect you want?"

Meaning: People often discuss and argue endlessly over actions without actually discussing why they want to do those actions in the first place. This leads to hours of wasted time, animosity between colleagues, and limitation of creative alternative actions to get to the outcome people wanted in the first place.

Use: When people are arguing with you, or others, over actions to take, or someone is generally being unclear in their explanation of what they want to do or why they want to do it.

A couple of examples from tech leaders:

One-liner: "Data beats emotions."—Sean Rad

Meaning: Don't use emotion in discussions to try to compel others to agree with you, and don't go around and around arguing over opinion. Get the data to make the decision; once you have facts, the answer is usually rather clear.

One-liner: "If you can't feed a team with two pizzas, it's too large." —Jeff Bezos

Meaning: As team size increases, productivity is a diminishing return. A primary reason for this is the bottleneck in the team. If teams become too large, they should be broken up into smaller teams.

[As an aside, I'm not sure Mr. Bezos has ever seen a hungry dev eat pizza, unless he wants team sizes of one or two.]

As you can see, the point of a one-liner is not to educate. It is assumed that people have an understanding of the new ways; however, their habits are driving them to behaviours, discussions, and decisions in line with the old way. These one-liners are a great way to snap them out of that, and engage System 2.

Process

- Use this on any change that is susceptible to this kind of cultural hurdle.
- Keep them short and easy to remember.
- Make them humorous or catchy (think marketing jingles).
- Make sure they're used in a fun/relaxed way.

People shouldn't be afraid to say them to reinforce the new way of doing things. Everyone should be ready to call each other out in a positive way to benefit the whole team. On that note, I even have a one-liner for that, from David Hurley (former Chief, Australian Defence Force):

"The standard you walk by is the standard you accept."

Implementation Tip Series

Pace Implementation Tip Series is a course of six videos by Peter Cronin giving a behind the scenes info on how we cause change rapidly and reliably. Each month Peter shares a further video from real world application tips to subscribers.

Subscribe free:
www. pacesoftwaredevelopment.com

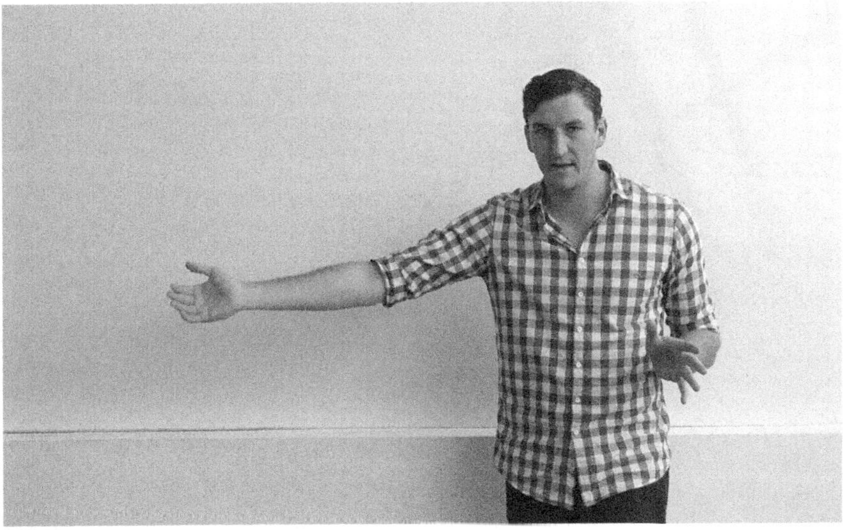

About the Author

Peter's passion is helping software companies scale without losing the magic.

He loves working with software companies because he finds them made up of the most interesting and innovative people on the planet.

Businesses Peter's ViAGO International team works with scale the start-up culture without the corporate burden that comes with growth. Their performance increases ahead of costs, founders work on the business rather than in the business, and the company, as a whole, drives toward the vision.

The team at ViAGO work with dev teams to reengineer operations in order to deliver higher quality code faster. They teach people techniques to make considered decisions instead of applying Band-Aid solutions, and they align individual efforts with the global strategy of the organisation.

Made in United States
Orlando, FL
04 November 2026

90202584R00108